AF762086

ENVIRONMENTAL SCIENCE, ENGINEERING AND TECHNOLOGY

HEAVY METAL SEDIMENTS

Environmental Science, Engineering and Technology

Additional books in this series can be found on Nova's website under the Series tab.

Additional E-books in this series can be found on Nova's website under the E-books tab.

ENVIRONMENTAL SCIENCE, ENGINEERING AND TECHNOLOGY

HEAVY METAL SEDIMENTS

LUCÍA H. SANZ
EDITOR

Nova Science Publishers, Inc.
New York

Copyright © 2012 by Nova Science Publishers, Inc.

All rights reserved. No part of this book may be reproduced, stored in a retrieval system or transmitted in any form or by any means: electronic, electrostatic, magnetic, tape, mechanical photocopying, recording or otherwise without the written permission of the Publisher.

For permission to use material from this book please contact us:
Telephone 631-231-7269; Fax 631-231-8175
Web Site: http://www.novapublishers.com

NOTICE TO THE READER

The Publisher has taken reasonable care in the preparation of this book, but makes no expressed or implied warranty of any kind and assumes no responsibility for any errors or omissions. No liability is assumed for incidental or consequential damages in connection with or arising out of information contained in this book. The Publisher shall not be liable for any special, consequential, or exemplary damages resulting, in whole or in part, from the readers' use of, or reliance upon, this material. Any parts of this book based on government reports are so indicated and copyright is claimed for those parts to the extent applicable to compilations of such works.

Independent verification should be sought for any data, advice or recommendations contained in this book. In addition, no responsibility is assumed by the publisher for any injury and/or damage to persons or property arising from any methods, products, instructions, ideas or otherwise contained in this publication.

This publication is designed to provide accurate and authoritative information with regard to the subject matter covered herein. It is sold with the clear understanding that the Publisher is not engaged in rendering legal or any other professional services. If legal or any other expert assistance is required, the services of a competent person should be sought. FROM A DECLARATION OF PARTICIPANTS JOINTLY ADOPTED BY A COMMITTEE OF THE AMERICAN BAR ASSOCIATION AND A COMMITTEE OF PUBLISHERS.

Additional color graphics may be available in the e-book version of this book.

Library of Congress Cataloging-in-Publication Data

Heavy metal sediments / editor, Lucma H. Sanz.
p. cm.
Includes bibliographical references and index.
ISBN 978-1-61470-690-8 (hardcover)
1. Soils--Heavy metal content. 2. Contaminated sediments. I. Sanz, Lucma H.
TD879.H4H433 2011
628.1'6832--dc23

2011024898

Published by Nova Science Publishers, Inc. † New York

CONTENTS

PREFACE

Metals are naturally present in our environment and trace amounts of metals are essential micronutrients for the growth of many marine organisms. However, excessive amounts of heavy metal can be toxic to marine life. In this book, the authors gather topical research from around the globe in the study of heavy metal sediments. Topics discussed in this compilation include the distribution of heavy metals and total petroleum hydrocarbons in soils near a petrochemical complex in China; heavy metals in coastal sea sediments of the eastern Adriatic Sea; environmental assessments of metal concentrations in select rivers, estuaries and coastal sediments; iron contamination in the Bizerte Lagoon of Tunisia and the correlation between millennial mining activities and heavy metal pollution in Spanish estuaries.

Chapter 1 - The influence of an industrialized environment on the accumulation of heavy metals and total petroleum hydrocarbons (TPH) in agricultural soils as well as the mobility and availability of metals from soil to vegetation were investigated. Seventy soil samples collected from the soil surface layers and horizons of five selected pedons in the vicinity area of a petrochemical complex in Guangzhou, China were analyzed for Zn, Cu, Pb, Cd, Hg, and As and TPH concentrations. The levels of metals in vegetation (lettuce, scallion, celery, tomato, carambola, wampee, and longan) were also evaluated. The horizontal and vertical variations of metals and TPH were assessed and the geographical information system (GIS)-based mapping techniques were applied to generate the spatial distribution maps. Results showed that the mean concentrations of heavy metals in top soils did not exceed the maximum allowable concentrations for the agricultural soil of China, with the exception of Hg. Significant differences were found depending on the land-use type. The concentrations of Cu, Pb, Cd, Hg, and As for vegetable field and orchard were significantly higher than for woodland, showing the metal concentrations in top soils strongly influenced by agricultural practices and soil management, while the TPH concentration in top soils was in the reverse order compared to the soil metal concentrations, in the order of woodland > orchard > vegetable field, implying strong influences of storage tank and pipeline leakage, accidental spills, and improper waste disposal practices. The concentrations of Zn, Cu, Pb, Cd, Hg, and As in five pedons, particularly in cultivated vegetable field and orchard, decreased with the soil depth, indicating these elements mainly originated from the anthropogenic sources, while the vertical variation of TPH concentrations showed increasing with the depth. The spatial distribution maps further allowed the identification of hot-spot areas with high contaminant concentrations. In addition, the similar spatial distribution pattern among metals reflected the similar sources. In comparison to the spatial distribution of TPH, metals showed very different patterns, corresponding to the differences in land-use types and vertical variation tendency. The major sources of elevated metals and TPH concentrations in soils were human activities, with metals mainly from wastewater irrigation and intensive application of sewage sludge and various agricultural chemicals and TPH mainly from storage tank and pipeline leakage, accidental spills, and improper waste disposal practices. When the modified European Community Bureau of Reference (BCR) three-step sequential extraction procedure was applied to determine the concentration of metal fractions in soils, the distribution of Zn, Cu, Pb, and Cd in four fractions significantly varied among soil samples, and 18.8% and 5.8% of vegetable and fruit samples for Cd and Pb, respectively, exceeded the maximum permissible levels for food-crops in China. The soil-to-vegetation transfer coefficients were in the order of Cd > Zn > Cu > Hg > As > Pb, suggesting Cd being the most mobile and available to plants among those metals studied. The principal component analysis (PCA) indicated that metal fractions and soil physicochemical properties (pH, organic matter, cation exchange capacity, clay content, and electrical conductivity) affected the metal uptake by plants. Effective measures should be taken to avoid or minimize further heavy metal contamination of soils and to remediate the contaminated areas in order to prevent pollutants from affecting human health through the agricultural products.

Chapter 2 - Polluted sediments present a threat to the health of the marine ecosystem and indirectly to the public health. In this work, more than 700 samples of the surface sea sediments have been collected in the eastern Adriatic coastal sea and analyzed for number of elements in effort to screen the whole area and to find the patterns of pollution pressure. In order to manage the large number of data, the GIS based database containing chemical element concentrations and descriptive data such as type of location have been produced. Results presented as GIS concentration distribution maps clearly show the presence of "hot-spots", which are investigated in details. In addition, the background values for measured chemical elements have been evaluated as a scientific basis for assessment of enrichment factors and quality criteria standards. The approach to this problem that author have been developed for the eastern Adriatic Sea is proposed to be applied elsewhere in particular to the Mediterranean Sea region.

Chapter 3 - Riverine, estuarine and coastal environment are important habitats for aquatic and marine life.However, long-term industrial developments with economic growth in coastal regions have resulted in environmental degradation, habitat loss and geomorphologic alteration in riverine, estuarine and coastal systems globally.Due to coastal development such as new industrial facilities and commercial port expansion, anthropogenic metals are introduced near the construction sites and dispersed and transported to the adjacent areas.Consequently, metal concentrations in riverine, estuarine and coastal sediments are one of the focused environmental concerns. Sediment quality reflects the long-term health status of a riverine, estuarine and coastal environment because it keeps the records of the historic development in the area.In this study, author summarize and report metal concentrations in the sediments of 43 selected sites around the world to expand the author's knowledge in understanding the environmental quality.The results from this study could be applicable to science-based policy formulation, and ecological restoration/ rehabilitation practices in an integrated coastal management (ICM) framework to meet the urgent need for integrated coastal management planning and environmental protection.

Chapter 4 - The depression occupied by Bizerte lagoon (northeast of Tunisia) corresponds to ante-Neogene solid mass ploughed up and eroded from Messinian. This area presents several wadis, in particular wadi Tinja (WT), natural gathered of the garaet Ichkeul, which is considered as the biggest surface of fresh water in Tunisia, and it is in permanent connection with the sea. The shoreline lagoon is urbanized and industrialized (e.g. steelworks of Menzel Bourguiba).

Many works show clearly the important role played by sediments in the pollutants geochemical cycle. Soussi (1981) indicates that the Bizerte lagoon sediments have primarily a muddy nature to a sandy-muddy containing organic matter.

Watkins and Pilkey (1973), show that heavy minerals present less than 1% of the sandy sediment of Bizerte bay. Yoshida (2004) supposes that one of the metallic pollution sources is the industrial activities, particularly in the vicinity of Menzel Bourguiba town; the other one is probably caused by pesticides pollution in the neighbouring agricultural zones and finally the urban rejections. (Srarfi, 2007) indicates that the maximum of organic pollution is recorded in the south-western part of the Bizerte lagoon. The mixture zones of fresh and salty waters constitute an important environment of marine chemical reactions, which are consecutive to the sudden change of environmental conditions (Ouakad 2007). This type of environment corresponds to estuary, delta or coastal lagoon with continental and marine influences.

Chapter 5 - The sedimentological, paleontological and geochemical analyses of two cores collected in the southwestern Spanish coast permits to delimitate the effects of different historical mining periods on the heavy metal contents of estuarine sediments. These effects are stronger in the Huelva estuary, with a first period of low to moderate pollution during Roman times (~2000 yr BP) and very high concentrations of heavy metals derived from recent mining activities (19th-20th centuries). On the contrary, sediments remain unpolluted or present very low metal contents in the external part of the Guadalquivir estuary during the last 2400 years, with absolute concentrations of Zn, Cu or Pb associated clearly to the grain size distribution.

Chapter 6 - The *Mar Menor* lagoon is one of the most unusual ecosystems in the Mediterranean area. Many economic and industrial activities concur in the zone and the combined impacts of mining, agriculture and urban development in the surroundings of the lagoon during recent decades have affected the ecosystem. In this chapter author review the impacts of mining activities on the lagoon, and report data on contamination levels and the effects generated in the lagoon ecosystem. The discharge of wastewaters from mineral refining processes appears to be the most important source of heavy metal inputs to the lagoon. Efforts to decrease these inputs must be made at the headwaters of streams where hundred of hectares of tailings remain unprotected against erosion.

Chapter 7 - With urbanization and economic development in coastal area, metal pollution in coastal environment has been a problem. Estuaries and coastal intertidal zone are important habitats for aquatic and marine life. In the meantime, sediment is a repository of contaminants and records the pollution history. Sediment quality reflects the long-term health status of an estuarine or coastal system and can be evaluated by sediment quality guidelines/criteria, metal enrichment factors and other indicators. In this study, author summarize the results mainly from the author's previous studies in metal pollution in sediments in urban estuarine and coastal systems in US and China, including New York Harbor and Yangtze River estuary in metropolitan areas and Xiamen Bay and Jiaozhou Bay. The information from this study expands the author's knowledge in understanding metal pollution in urban coastal systems and assessing environmental quality impacted by industrialization and economic development.

In: Heavy Metal Sediments
Editor: Lucía H. Sanz

ISBN: 978-1-61470-690-8
© 2012 Nova Science Publishers, Inc.

Chapter 1

DISTRIBUTION OF HEAVY METALS AND TOTAL PETROLEUM HYDROCARBONS IN SOILS NEAR A PETROCHEMICAL COMPLEX IN GUANGZHOU, CHINA

Junhui Li and Hojae Shim*
Department of Civil and Environmental Engineering,
Faculty of Science and Technology, University of Macau,
Macau SAR, China

ABSTRACT

The influence of an industrialized environment on the accumulation of heavy metals and total petroleum hydrocarbons (TPH) in agricultural soils as well as the mobility and availability of metals from soil to vegetation were investigated. Seventy soil samples collected from the soil surface layers and horizons of five selected pedons in the vicinity area of a petrochemical complex in Guangzhou, China were analyzed for Zn, Cu, Pb, Cd, Hg, and As and TPH concentrations. The levels of metals in vegetation (lettuce, scallion, celery, tomato, carambola, wampee, and longan) were also evaluated. The horizontal and vertical variations of

---

* Corresponding author. Tel.: 853-83974374; Fax: 853-28838314; E-mail: hjshim@umac.mo (Dr. H. Shim).

metals and TPH were assessed and the geographical information system (GIS)-based mapping techniques were applied to generate the spatial distribution maps. Results showed that the mean concentrations of heavy metals in top soils did not exceed the maximum allowable concentrations for the agricultural soil of China, with the exception of Hg. Significant differences were found depending on the land-use type. The concentrations of Cu, Pb, Cd, Hg, and As for vegetable field and orchard were significantly higher than for woodland, showing the metal concentrations in top soils strongly influenced by agricultural practices and soil management, while the TPH concentration in top soils was in the reverse order compared to the soil metal concentrations, in the order of woodland > orchard > vegetable field, implying strong influences of storage tank and pipeline leakage, accidental spills, and improper waste disposal practices. The concentrations of Zn, Cu, Pb, Cd, Hg, and As in five pedons, particularly in cultivated vegetable field and orchard, decreased with the soil depth, indicating these elements mainly originated from the anthropogenic sources, while the vertical variation of TPH concentrations showed increasing with the depth. The spatial distribution maps further allowed the identification of hot-spot areas with high contaminant concentrations. In addition, the similar spatial distribution pattern among metals reflected the similar sources. In comparison to the spatial distribution of TPH, metals showed very different patterns, corresponding to the differences in land-use types and vertical variation tendency. The major sources of elevated metals and TPH concentrations in soils were human activities, with metals mainly from wastewater irrigation and intensive application of sewage sludge and various agricultural chemicals and TPH mainly from storage tank and pipeline leakage, accidental spills, and improper waste disposal practices. When the modified European Community Bureau of Reference (BCR) three-step sequential extraction procedure was applied to determine the concentration of metal fractions in soils, the distribution of Zn, Cu, Pb, and Cd in four fractions significantly varied among soil samples, and 18.8% and 5.8% of vegetable and fruit samples for Cd and Pb, respectively, exceeded the maximum permissible levels for food-crops in China. The soil-to-vegetation transfer coefficients were in the order of Cd > Zn > Cu > Hg > As > Pb, suggesting Cd being the most mobile and available to plants among those metals studied. The principal component analysis (PCA) indicated that metal fractions and soil physicochemical properties (pH, organic matter, cation exchange capacity, clay content, and electrical conductivity) affected the metal uptake by plants. Effective measures should be taken to avoid or minimize further heavy metal contamination of soils and to remediate the contaminated areas in order to prevent pollutants from affecting human health through the agricultural products.

Keywords: Bioavailability, Guangzhou, Heavy metals, Petrochemical complex, Spatial distribution, TPH

1. INTRODUCTION

In modern societies, the introduction of mitigation measures for pollution abatement has progressively reduced the amount of chemicals released to the environment. However, the industrial sector is still being an important source of environmental contaminants [1]. Petrochemical industries have been identified as important emission sources of a wide variety of pollutants [1, 2]. Significant amounts of organic contaminants, such as benzene, toluene, ethyl-benzene, and xylenes (BTEX), polycyclic aromatic hydrocarbons (PAHs), polychlorinated biphenyls (PCBs), together with other hydrocarbons, including total petroleum hydrocarbons (TPH), have been detected in surroundings near oil refineries [2-6]. Most heavy metal emissions are the result of industrial activity, and one of the identified potential sources is the oil industry [7]. The release of metals such as vanadium (V), arsenic (As), and chromium (Cr) may occur in either direct stack emissions from refineries or indirect emissions from the combustion of hydrocarbon fuels [1, 2, 7].

Contaminants accumulation in agricultural soils is of increasing concern worldwide, particularly in China with the rapid development of industrialize-tion and urbanization [8-12]. Plants grown in polluted soils can accumulate potentially toxic elements at high concentrations and cause a serious risk to human health when the plant-based food stuffs are consumed [4, 9, 13-15]. In general, sources of heavy metals in agricultural soils include natural occurrence derived from parent materials, as well as such anthropogenic activities as mining, smelting, waste disposal, urban effluent, vehicle exhausts, sewage sludge generation, and applications of pesticides and fertilizers [16, 17]. Many cases of severe metal pollution in soils by anthropogenic sources have been reported for areas surrounding industries [16, 18-20]. Soil contaminant concentrations were found to decrease with increasing distance from the industrial area [10, 11, 21, 22], and in most cases, contaminant concentrations in top soils adjacent to the industrial areas are significantly higher than background levels [21-23]. Although the atmospheric deposition could lead to a serious soil contamination in the vicinity of industrial facilities, it would not necessarily mean the main reason for high soil metal concentrations in surrounding areas. The evidence for attributing the origin of soil pollution to a particular emitter is usually indirect and based on the spatial gradients of decreasing pollutant concentrations with distance from the suspected source [18]. Moreover, the vertical distribution of element concentrations with depth is a means of identifying geogenic and anthropogenic sources to some extent [19, 20].

Many studies had carried out about the environmental quality and its effect on human health around petrochemical areas. Several studies had found the excess cancer risk for communities surrounding the petroleum refinery plant [24-27]. Studies had also shown that the residential petrochemical air pollution could affect the outcome of pregnancies, such as sex ratios at birth [25], the excess risk of delivering low-birth weight babies at term [26], and the higher prevalence of delivering preterm birth infants [27].

Research on the mechanisms of metal uptake by crops from contaminated soils is of increasing concern worldwide due to the potential effects on human health [9, 28, 29]. High concentrations of metals (Cu, Cd, and Pb) in fruits and vegetables were shown to be related to the high prevalence of upper gastrointestinal cancer [30, 31]. The main sources of metals for plants are their growth media (e.g., soil, nutrient solutions) from which metals are taken up by root [32, 33]. Most of the previous studies have shown that the increasing concentrations of metals in soil may increase plant uptake, and the uptake and accumulation of metals in plants are influenced by a number of factors such as availability of metals in soil, plant species, plant age, climate, and atmospheric depositions [28, 31, 33-35]. Consequently, the metal availability to plants is quite important when assessing the effect of soil contamination on plant metal uptake [35, 36]. The availability of certain metals reportedly depends on the physicochemical properties of soil such as pH, organic matter (OM) or cation exchange capacity (CEC) and on the distribution of metals among various soil fractions [28, 34, 35, 37].

Since 1973, one of the largest petrochemical complexes in China has been located in Huangpu District, Guangzhou City, China, with the petroleum refinery together with the petrochemical industry within the area. The refinery has the capacity to process up to 13 million tons of crude oil and produce 200,000 tons of ethylene per year. Any emissions from the oil refineries may directly affect residents living in the area. Petrochemical refineries produce a wide variety of petroleum and chemical products such as fuels, lubricant oils, natural gas, petroleum coke, and asphalt, and numerous pollutants are discharged into the surroundings. The majority of residential areas in this study are located downwind of the petrochemical complex, increasing the potential health risk. In recent years, public concern over possible adverse health effects for the population living near this industrial complex has increased, especially toward the accidents. The most recent accidental explosion occurred on March 18, 2010. There have been about 10 accidental explosions or burning accidents since the operation of the petrochemical complex.

However, little information is available about the contaminant concentrations in agricultural soils around petrochemical complex at present. In response to the concern of the local population to these facilities, this study has recently been initiated, focused on determining the current levels of Zn, Cu, Pb, Cd, Hg, As, and TPH, establishing the health risks for the population living in the neighborhood of the petrochemical complex, elaborating the spatial distribution maps of the contaminants in agricultural soils around the complex, as well as identifying the pollution sources. At the same time, the levels of metal contamination in edible portions of some vegetables and fruits were quantified and the modified European Community Bureau of Reference (BCR)-operational speciation of metals (Cu, Zn, Cd, and Pb) in vegetable field and orchard soil samples was analyzed to evaluate the availability of metals from soil to plants. Soil organic matter, pH, CEC, electrical conductivity (EC), and clay content were also determined to investigate their possible influences on metal uptake by crops in agricultural soil, and the metal transfer from soil was estimated. Finally, the principal component analysis (PCA) was employed to investigate the relationships between metal content of plants and agricultural soil in the petrochemical area.

2. Materials and Methods

2.1. Study Area

The study area is located immediately south of the Pearl River and near a petrochemical complex in the Huangpu District of Guangzhou City (Latitude 22°26′ to 23°56′ N, Longitude 112°57′ to 114°03′ E), the capital of Guangdong Province in South China (Figure 1). The area is located in the surrounding uplands with a granitic soil parent material, and the soil types are Plinthosols and Anthrosols according to the World Reference Base soil classification [29]. The area shows a typical subtropical monsoonal climate, with the annual average temperature of 21.4-21.8℃ and the average rainfall of 1,689.3-1,876.5 mm [38], which is very suitable for the growth of most subtropical fruits and vegetables [12]. Winds predominantly originate from the north in winter and east to southeast in summer (Figure 2). The frequencies of prevailing wind directions in winter and summer are similar. Two canals, serving as the main irrigation water sources of the arable land, flow through the study area and become the recipients of municipal sewage and industrial wastewater before directly discharged into the Pearl River. The agricultural activity is centered on vegetables and fruits.

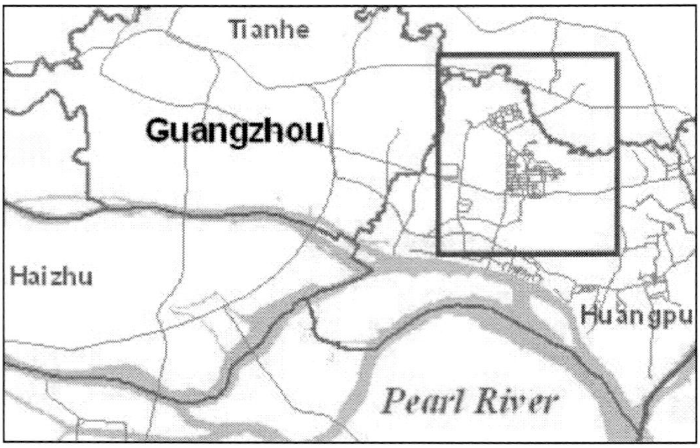

Figure 1. Location map of the study area.

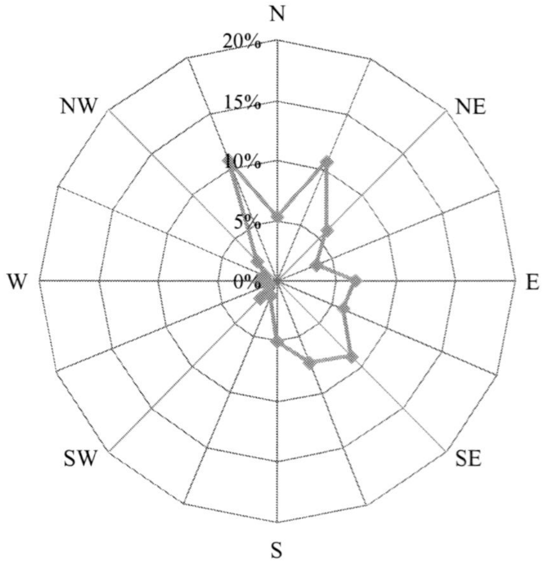

Figure 2. Wind frequency in directions to Guangzhou.

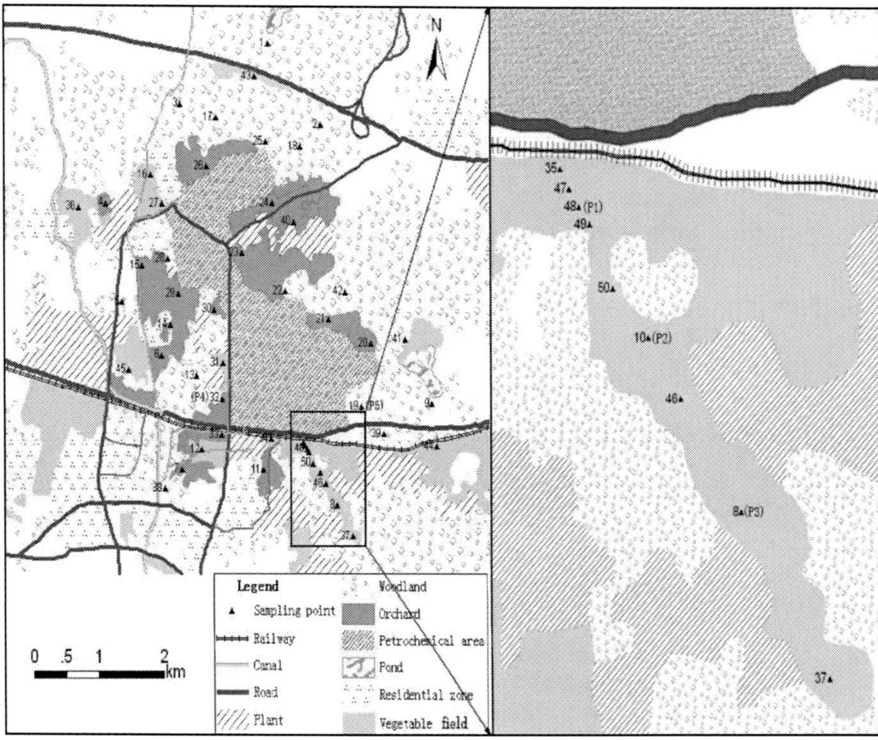

Figure 3. Soil sampling points in the study area.

2.2. Sampling

A total of 70 soil samples was collected from 50 different sites around the petrochemical complex in January, 2007 (Figure 3). The sites were chosen primarily to reflect both increasing distances (100, 500, 900, and 1,300 m) from the boundary of the petrochemical complex and different land-use types, including vegetable field, orchard, and woodland. The composite topsoil samples (0-20 cm) were obtained by mixing five random sub-samples within 25 m^2 in each sampling site. A transect at the south vegetable filed area was chosen and the topsoil samples were collected at 100, 150, 200, 250, 400, 550, 700, 1,000, and 1,300 m away from the boundary of the petrochemical complex. Five pedons (P1, P2, P3, P4, and P5) located at sampling points No 48, 10, 8, 32, and 19, respectively, were dug and soil samples collected at 0-10, 10-20, 20-40, 40-60, and 60-80 cm depth individually. All soil samples were collected using hand auger. Seven common vegetative species of lettuce

(*Lactuca sativa Linn. var. ramosa Hort.*), scallion (*Allium fistulosum Linn., Liliaceae*), celery (*Apium graveliens L., Umbelliferae*), tomato (*Lycopersicon esculentum, Solanaceae*), carambola (*Averrhoa carambola Linn., Oxalidaceae*), wampee (*Clausena lansium (Lour.) Skeels, Rutaceae*), and longan (*Dimocarpus longan Lour., Sapindaceae*) were randomly collected on the sites where soil samples were taken. Each plant sample was composed of 3-5 subsamples. Since sometimes more than one kind of plant was grown on the same site, plants were sampled individually. Both soil and plant samples were stored in polyethylene bags before use.

2.3. Chemicals

The TPH stock solution was prepared using the 1:1 ratio (v/v) of unleaded gasoline (purity, 99%) and diesel (purity, 99%) purchased from the Caltex Company in Macau SAR, China. Dichloromethane (purity, 99%) and acetone (purity, 99%) were purchased from the International Laboratory (USA).

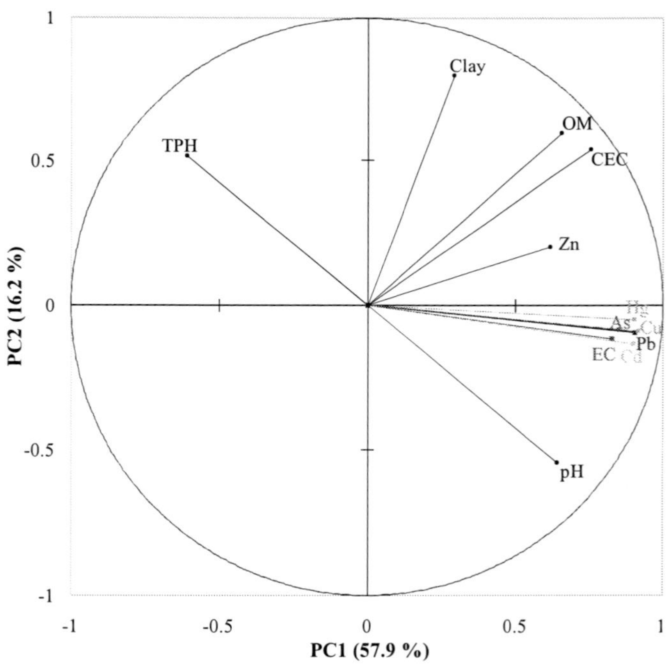

Figure 4. Plot of the first two principal components obtained from the PCA of soil TPH, soil metals, and selected soil parameters.

Table 1. The modified BCR sequential extraction procedure

Step	Extraction procedure	Operationally defined phase
i	40 mL of 0.11 mol L^{-1} HOAc per 1 g of dry soil sample shaken for 16 h on mechanical shaker at 25℃.	HOAc extractable (water soluble, exchangeable and bound to carbonate)
ii	40 mL of 0.5 mol L^{-1} $NH_2OH{\cdot}HCl$ (adjusted to pH=1.5 with HNO_3) added to residue and shaken for 16 h at 25℃.	Reducible (bound to Fe and Mn oxides)
iv	10 mL of 8.8 mol L^{-1} H_2O_2 added to residue and digested for 1 h at 25℃, and then for 1 h at 85℃ in a water bath with a second volume of H_2O_2. Next the solution evaporated to a few mL. 50 ml of 1 mol L^{-1} NH_4OAc (adjusted to pH=2.0 with HNO_3) added to residue and shaken for 16 h at 25℃.	Oxidizable (bound to organic matter and sulfides)
iv	HNO_3-HF-$HClO_4$ digestion, the same procedure as the total metal determination.	Residual (bound to silicate minerals)

2.4. Analytical Methods

Soil samples were first air-dried, crushed to pass through a ten mesh (<2 mm) nylon sieve, and homogenized. The sub-samples were then ground with an agate grinder to go through a 100 mesh (<150 μm) nylon sieve. Then, soil samples were taken to perform the sequential extraction procedure in order to assess the metal fractions, especially the available pool. The soil properties including pH, EC, CEC, OM, and clay content were also determined according to the standard methods [39], i.e., OM was determined by the wet oxidation at 185℃ in a mixture of potassium dichromate and concentrated sulfuric acid, soil pH and EC were measured in a 1:2.5 (w/v) ratio of soil to distilled water suspension, CEC was measured after extraction with 1.0 mol L^{-1} ammonium acetate (pH 7.0), and clay content was determined using the pipette method. Soil Hg was digested in a mixture of H_2SO_4, HNO_3, and $KMnO_4$, and soil As was digested with HCl and HNO_3. All other soil metals were digested in a mixture of concentrated HNO_3, HF, and $HClO_4$ in a polyvinyl fluoride crucible [29, 40].

The chemical partitioning of Cd, Pb, Cu, and Zn in soil samples was performed using the modified BCR procedure [41]. The sequential extraction method operationally defines the metals in four chemical forms: HOAc extractable, reducible, oxidizable, and residual fractions. The three-step sequential extraction procedure was carried out step-by-step as summarized in Table 1. The metal recovery from the sequential extraction analysis was assessed by comparing the sum of all fractions with the total metal concentrations for each soil sample.

The edible parts of each plant were simultaneously weighed after washing with Mill-Q water and blotting dry. All vegetative samples were dried in an oven at 65℃ to the constant weight, weighed and ground into fine powder by a stainless steel mill. Then, 1.0 g sub-samples were digested in a mixture of concentrated HNO_3 and $HClO_4$ for Cd, Pb, Cu, and Zn, while As was digested with HCl and HNO_3. Hg was digested using a mixture of V_2O_5, H_2SO_4, and HNO_3 [29, 40].

Cu and Zn of the above resulting digests were determined by a flame atomic absorption spectrophotometer (Hitachi Z-5000) and Cd and Pb by a graphite furnace atomic absorption spectrophotometer (GFAAS-AA800, PerkinElmer Inc.), while Hg was measured by the cold vapor atomic absorption (F732-V, Shanghai) and As by the hydride generation atomic fluorescence spectrometry (HG-AFS 230, Beijing). The quality assurance and quality control of metal analyses were assessed in duplicates using duplicates, reagent blanks, and standard reference materials [42]. The recoveries of analyzed metals in standard reference materials were within ±10% of the recommended values, and the relative standard deviation of duplicate measurements was less than 10%. The recovery values obtained by proportioning the results obtained by the BCR procedure to those of the total digestion were also found satisfactory.

The accelerated solvent extraction (ASE) was applied for the quantitative extraction of TPH from soil and the GC-FID-based TPH-determination method was used to analyze the extracts. TPH (4,000 mg kg^{-1}) as the internal standard was added to the sample vials to determine the recovery efficiency (96-102%).

All extractions were performed using the Dionex (Sunnyvale, CA, USA) ASE 200 system. For the extraction, 5 g soil (<2 mm fraction) mixed (5:1) with the ASE Prep DE (diatomaceous earth) were weighted out into the extraction cell (11-mL stainless steel cells). The glass fiber filters (Restek, Bellefonte, PA, USA) were placed at the outlet of each cell prior to being loaded with samples. The optimized condition [43, 44] was employed for the extraction, at 175℃ with dichloromethane-acetone (1:1, v/v) with 8 min heat-up time and 5 min static time with the pressure of 1,500 psi nitrogen gas for all samples (1 psi=56,894.76 Pa). The flush volume was 70% and the purge time was 60 s with 150 psi. The extracts were collected in pre-cleaned 40-mL glass vials (I-Chem, New Castle, DE, USA).

Quantification of the extracts was performed on a gas chromatograph (Agilent, 6890N, Agilent Technologies Co., Ltd, China) equipped with a flame ionization detector (FID) and a capillary column (Agilent HP-5; 30 m × 0.53 μm I.D. with a stationary-phase film thickness of 0.88 μm). One microliter of the liquid sample was injected by the autosampler injector (7638 Series, Agilent Technologies Co., Ltd, China) equipped with a tapered microsyringe (Hamilton 5181-1267, Hamilton Company, USA). The temperature of injector and detector was 230℃ and 280℃, respectively. The temperature program for column was with the initial temperature 40℃ (hold for 2 min) and incrementally increased (12℃/min) to 280℃ (hold for 20 min) [45].

2.5. Statistical Analysis and GIS Mapping

All statistical analyses were performed using the Microsoft Excel 2003 and SPSS V13.0 for Windows. The analysis of variance was used to examine the statistical significant differences in the mean metal concentration among soil groups. A probability level of $p<0.05$ was considered as a significant difference. The Principal Component Analysis (PCA) based on the correlation matrix was performed with the XLStat-Pro 7.5.2 software, used as a Microsoft Excel plug-in. The map of spatial distribution of heavy metal and TPH concentrations was generated by kriging interpolating data from 50 surface soils by using the ArcGIS V9.0 (ESRI Corporation). The values of metal concentration were grouped into eight categories by using automatic break criterion of the Arcmap, one of the ArcGIS programs, and the values in different groups were presented with different colors.

The transfer coefficients (TC) of Zn, Cu, Pb, Cd, Hg, and As from soils to plants (edible parts) were calculated according to the following formula:

$$TC = \frac{P\ (\text{mg kg}^{-1}\ \text{fresh weight})}{S\ (\text{mg kg}^{-1}\ \text{dry weight})}$$

where P represents the metal concentration in plant tissue and S is the total metal concentration in soil where the plant grows.

Table 2. Comparison of topsoil properties for three different land-use types

	Vegetable field (n=20)		Orchard (n=14)		Woodland (n=16)	
	Range	Mean±SD	Range	Mean±SD	Range	Mean±SD
pH (H_2O)	5.05-7.74	6.22±0.81c	3.32-7.78	4.72±1.30b	2.94-6.56	3.79±1.09a
OM (g kg^{-1})	12.60-46.56	25.99±9.06a	11.57-44.71	22.90±9.93a	12.4-37.41	23.72±7.28a
CEC (cmol kg^{-1})	3.79-12.89	7.92±2.45b	3.06-10.91	6.36±2.47a	3.44-9.37	5.98±1.58a
EC (μs cm^{-1})	234-932	480±193b	65-298	174±70a	105-206	152±31a
Clay (<2 μm, %)	9.34-40.50	28.06±8.54a	13.90-37.43	26.30±7.27a	16.69-37.43	28.21±5.35a

Results are presented as arithematic mean±SD.
Mean±SD within a row for a certain land-use type followed by different letters are significantly different at the 0.05 level.
n is the number of samples.

Table 3. Heavy metal and TPH concentrations in topsoils (mg kg^{-1})

	Zn	Cu	Pb	Cd	Hg	As	TPH
Range	19.6-386.1	3.6-294.3	25.4-182.1	0.02-1.03	0.028-2.4	0.012-15.5	1,179.3-6,354.9
Mean ± SD	158.6±82.2	38.8±48.7	69.4±37.4	0.23±0.21	0.61±0.64	4.6±3.6	2,676.6±1,182.9
Median	146.3	17.5	60.4	0.16	0.45	4.5	2521.7
Coefficient of variation (%)	51.8	125.4	53.8	88.9	104.0	77.8	44.2
Background values of natural soil in Guangzhou [a]	58.1	13.6	42.88	0.14	0.157	17.4	-
Maximum allowable concentrations in agricultural soil of China[b]	200	50	250	0.3	0.3	40	-

[a] Guangdong Province Environmental Monitoring Center, 1990.
[b] National Environmental Protection Agency of China, 1995.

3. Results and Discussion

3.1. Basic Soil Physico-Chemical Properties

The pH of topsoils ranged from strongly acidic (pH 2.9) to slightly alkaline (pH 7.8) (Table 2). Soil pH varied significantly between the three land-use types ($p<0.05$). The mean topsoil pH was the greatest in vegetable field (6.22), followed by orchard (4.72) and woodland (3.79). The significant differences between three land-use types were likely due to the application of lime and various fertilizers on vegetable field and orchard.

The soil CEC and EC values ranged from 3.06 to 12.89 cmol kg^{-1} and 65 to 932 μs cm^{-1}, respectively, with the average of 7.92 cmol kg^{-1} and 480 μs cm^{-1}, respectively (Table 2). The CEC was low for all soils, especially for woodland. The significant differences of CEC and EC in topsoils were found between land-use types. The mean soil CEC and EC were significantly higher in vegetable field compared with the other two land-use types. Woodland topsoils showed the lowest values of CEC and EC.

The OM and clay contents in topsoils ranged from 11.6 to 46.6 mg kg^{-1} and 9.3% to 40.5%, respectively, with the average of 24.4 mg kg^{-1} and 27.6%, respectively (Table 2). The mean soil OM and clay contents were not significantly different among the three land-use types.

3.2. Heavy Metal and TPH Concentrations in Soils

The concentrations of heavy metals and TPH in agricultural topsoils in the proximity of the petrochemical complex showed that the Zn, Cu, Pb, Cd, Hg, As, and TPH concentrations ranged from 19.6 to 386.1, from 3.6 to 294.3, from 25.4 to 182.1, from 0.020 to 1.03, from 0.028 to 2.4, from 0.012 to 15.5, and from 1,179.3 to 6,354.9 mg kg^{-1}, with the average of 158.6, 38.8, 69.4, 0.23, 0.61, 4.6, and 2,676.6 mg kg^{-1}, respectively (Table 3). In comparison with the background elemental values of natural soil in Guangzhou [46], the mean concentrations of Zn, Cu, Pb, Cd, and Hg in topsoils were higher, while the As concentration was much lower. The percentage of studied soils with the concentrations of Zn, Cu, Pb, Cd, and Hg larger than the background value was 92%, 64%, 64%, 58%, and 64%, respectively. The mean Zn, Cu, Pb, and Cd concentrations in topsoils were lower than those in urban soils in this area [38]. In accordance with the Environmental Quality Standard for Soils [42], the mean concentrations of Zn, Cu, Pb, Cd, and As in topsoils were lower than the maximum allowable concentrations in the agricultural soil of China with the exception of Hg (Table 3), whereas the percentages of sample topsoils exceeding the maximum allowable concentrations were 30% for Zn, 28% for Cu, 22% for Cd, and 52% for Hg, respectively. The Pb and Hg concentrations in topsoils were higher and the Cd and As concentrations were roughly similar compared to soils around a petrochemical complex in Spain [4]. Compared with other industrial areas, these metal concentrations would be in the lower part of the range [16, 21, 23], while the TPH concentration is higher compared to other petrochemical areas [22, 47-50].

Results also indicated that the soil Zn, Pb, Cd, Hg, As, and TPH concentrations varied among three land-use types (Table 4). The average concentration of these metals in topsoils was in the order of vegetable field > orchard > woodland, while in case of Cu, the average concentration in the order of orchard > vegetable field > woodland. In comparison, the average TPH concentration in topsoils was in the reverse order, i.e., woodland > orchard > vegetable field. Furthermore, statistically significant differences between land-use types were found for the topsoil heavy metal and TPH levels, with the exception of Zn. The soil Cu, Pb, Cd, Hg, and As concentrations for vegetable field and orchard were significantly higher than for woodland. Phosphatic fertilizers had been widely regarded as a potential source of Cd, As, and Hg contamination of agricultural soils, and the applications of manures, sewage sludge, and pesticides found a more significant non-point source of Cd, Cu, Pb, and Zn in soils [16, 17]. Results showed that the heavy metal levels in topsoils are strongly influenced by such intensive farming of vegetables growing all year round and the same results also reported by others [12, 51]. The soil TPH concentrations for woodland and orchard were significantly higher than for vegetable field, implying a possible result from air emissions, storage tank and pipeline leakage, accidental spills, and improper waste disposal practices. Concentrations of trace elements with the anthropogenic sources in soil had been expected highly variable from one sampling point to another because their concentrations depended on the distances between sampling points and pollution sources [52]. The large standard deviation and high coefficient of variation (CV) of their concentrations (Table 3) accounted for the strong variability as expected.

The correlation analysis results (Table 5) showed that there were significant correlations among Zn, Cu, Pb, Cd, Hg, and As concentrations in surface soils, except for Cu against Zn. The significant positive correlations were caused by the physicochemical features of elements and geochemical processes. This also indicated that the studied metals were derived mainly from the same contaminating sources. Many previous studies also found the similar results [22, 23, 53, 54].

Moreover, there existed a significant correlation between Zn, Cu, Pb, Cd, Hg, and As concentrations and soil organic matter (Table 4), implying that the organic manures, such as pig and poultry litter, urban wastes, sewage sludge, etc., might be one of the main sources of these heavy metal pollutants, and the organic matter might have helped retain metals that entered the soils because of the high affinity between metals and organic matter [17, 55].

The PCA (Figure 4) was performed with the TPH concentration, the total concentrations of each metal, and the selected soil properties, in order to analyze the relationships among these indices. The first principal component, with the relative inertia of 57.9%, was positively determined by metals and EC but negatively determined by TPH. The second principal component, with the relative inertia of 16.2%, was positively determined by clay, OM, and CEC but negatively determined by pH. The TPH concentration was shown almost in a straight line toward pH, indicating negatively significant correlations between them. In addition, the TPH concentration was negatively correlated with each metal, similar to the correlation analysis results (Table 5). Since such human activities as wastewater irrigation and intensive application of sewage sludge and various agricultural chemicals are the main reasons for the heavy metal elevation in soils and the source of TPH is different from the metals, it was considered that the oil leakage was possibly the main source of TPH accumulated in soil.

Table 4. Comparison of heavy metals and TPH concentrations (mg kg^{-1}) in topsoils for different land-use types

	Vegetable field (n=20)	Orchard (n=14)	Woodland (n=16)
Zn	176.68±80.97a	175.29±86.97a	121.34±70.78a
Cu	51.95±36.96a	52.91±73.71a	10.15±4.14b
Pb	97.13±34.68a	66.20±31.37b	37.66±8.88c
Cd	0.37±0.25a	0.19±0.10a	0.094±0.042b
Hg	0.94±0.55a	0.72±0.75a	0.10±0.094b
As	6.40±2.85a	5.32±3.95a	1.66±1.91b
TPH	1,934.3±596.2b	2,886.6±1,251.3a	3,420.8±1,181.9a

Results are presented as arithematic mean±SD.

Mean±SD within a row for a certain land use type followed by different letters are significantly different at the 0.05 level.

n is the number of samples.

Table 5. Pearson correlation coefficient matrix for the topsoil variables (n=50)

n=50	Zn	Cu	Pb	Cd	Hg	As	TPH	OM	pH	CEC	EC	Clay
Zn	1											
Cu	0.246	1										
Pb	0.427**	0.537**	1									
Cd	0.476**	0.435**	0.779**	1								
Hg	0.380**	0.419**	0.741**	0.559**	1							
As	0.441**	0.458**	0.758**	0.584**	0.822**	1						
TPH	-0.162	-0.363**	0.480**	-0.475**	-0.416**	-0.373**	1					
OM	0.436**	0.327*	0.525**	0.552**	0.498**	0.551**	-0.24	1				
pH	0.294*	0.498**	0.600**	0.459**	0.491**	0.524**	-0.475**	0.099	1			
CEC	0.454**	0.341*	0.646**	0.617**	0.589**	0.645**	-0.306*	0.842**	0.215	1		
EC	0.332*	0.305*	0.641**	0.689**	0.415**	0.488**	-0.533**	0.344*	0.534**	0.581**	1	
Clay	0.206	0.017	0.124	0.046	0.199	0.173	0.215	0.470**	-0.065	0.573**	0.218	1
Dist	-0.319*	-0.078	0.013	0.025	0.108	0.074	-0.178	-0.156	0.27	-0.166	0.022	-0.338*

*Correlation significant at $p<0.05$.

**Correlation significant at $p<0.01$.

Dist =distance (m) from the petrochemical complex.

n=number of samples.

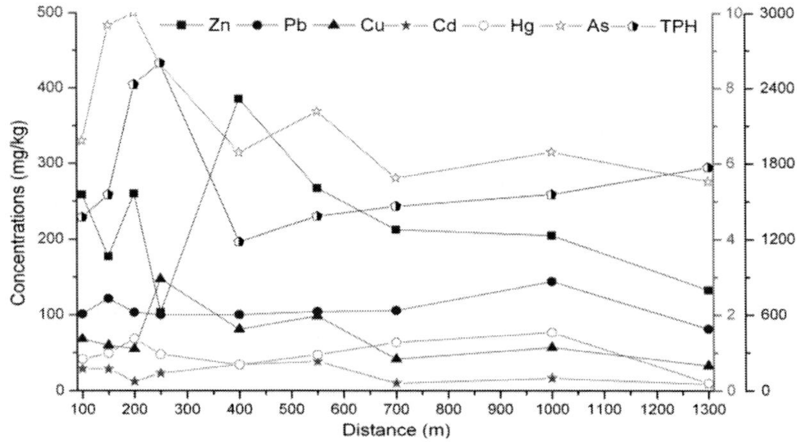

Figure 5. Concentrations of Zn, Cu, Pb, Cd, Hg, As, and TPH in topsoils (0-20 cm) vs. distance from the petrochemical complex.

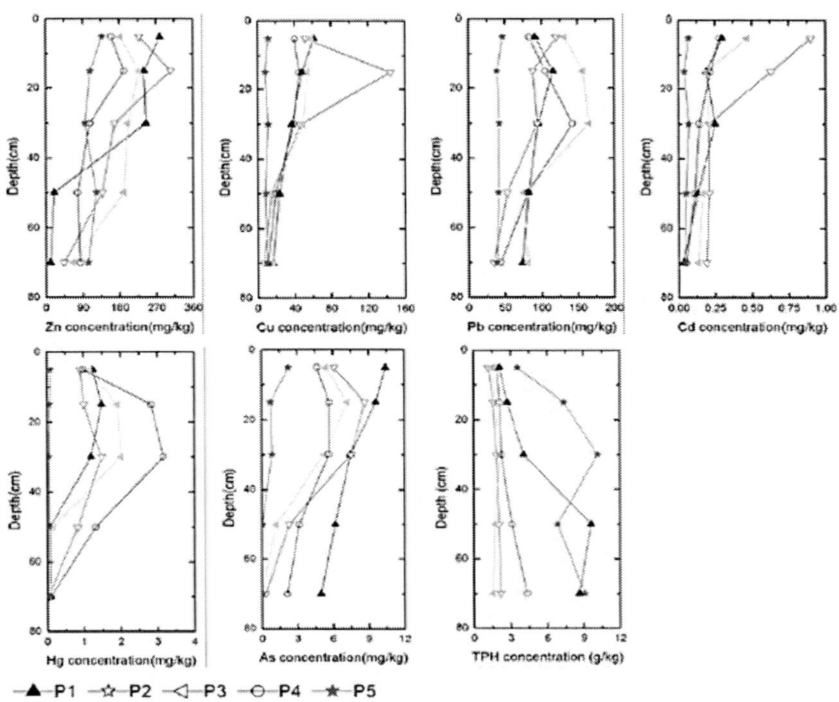

Figure 6. Vertical variations of Zn, Cu, Pb, Cd, Hg, As, and TPH concentration in soil pedons.

3.3. Horizontal and Vertical Variability of Heavy Metals and TPH in Soils

There were no distinct gradients of Zn, Cu, Pb, Cd, Hg, As, and TPH concentrations with distance (within 1,300 m) from the petrochemical complex (Figure 5). This suggested little influence of petroleum air emission on the soil heavy metal accumulation. A significant negative correlation was observed for the Zn concentration in topsoils versus the distance from the complex (Table 5), suggesting that Zn in soils might be related to the air emission from the complex. The concentrations of Cu, Pb, Cd, Hg, and As in soils, however, were not significantly correlated with the distance, further supporting that little influence of the air emission from the complex on the soil heavy metal accumulation.

The vertical variation of Zn, Cu, Pb, Cd, Hg, and As concentrations in five pedons (Figure 6) showed that these elemental contents had a systematical decreasing tendency with the increase of depth. This trend supported that the anthropogenic heavy metal contamination is mainly concentrated in the surface. In the woodland pedon (P5), the concentrations of heavy metals were the lowest and showed little variation with the depth increase, compared to vegetable field (P1, P2, and P3) and orchard (P4), indicating the anthropogenic sources with the least influence on the heavy metals in woodland soils and these elements predominantly originated from the soil-forming parent materials and the atmospheric deposition. The concentration of Hg in pedons for vegetable field and orchard initially increased with depth, and then decreased sharply, finally less than the detection limit in the deep horizons. Previously Hg was reported combined with the abundant organic matter in the surface soils [55], which would help prevent the further downward movement. Similarly, Svendsen et al. [22] reported Hg was mainly concentrated in the surface soil within the profile in the vicinity of a Zn smelter. The concentration of Cd also decreased with depth. The trend of Hg and Cd concentrations with depth further supported that the two elements were mainly characterized by the anthropogenic sources. The concentrations of Zn and As also exhibited a large variability within and among the five pedons. In surface layers, the highest Zn and As concentrations were in soils from vegetable fields, followed by orchard, and the soil from woodland was the lowest. Nevertheless, the adverse trend was observed for the Zn concentration in the bottom layer (60-80 cm), which could be explained by the Zn gradient of the original deposition of the parent material. The concentration of Cu in 0-20 cm depth showed a marked difference, whereas below the 50 cm depth, the concentrations in all pedons showed little difference, reflecting fertilization, irrigation, and pesticide application contributed to the Cu elevation. However, the vertical variation of TPH concentrations in five pedons (Figure 6) showed that the TPH content had a systematical increasing tendency with the increase of depth. The TPH concentrations in vegetable field (P1, P2, and P3) were the lowest and showed little variation with the depth increase, compared to woodland pedon (P5) and orchard (P4). For the vegetable field, the TPH concentrations in pedons 1 and 2 increased with depth, while for pedon 3, initially increased with depth and then decreased. Similar to pedons 1 and 2 in the vegetable field, the TPH concentrations in pedon 4 (orchard) and pedon 5 (woodland) showed the increasing trend with depth. The TPH compounds showed a great tendency to leach and migrate into the subsurface soil or groundwater, potentially impacting groundwater.

3.4. Spatial Distribution of Heavy Metals and TPH in Topsoils

The spatial distribution of contaminant concentrations is a useful visual aid to assess the possible sources of enrichment and to identify hot-spot area with the high contaminant concentration [19, 52-54]. The spatial distribution of Zn, Cu, Pb, Cd, Hg, As, and TPH in soils was analyzed using GIS methods. The contaminant concentrations were interpolated with the kriging method. The maps of Zn, Cu, Pb, Cd, Hg, As, and TPH are presented in Figure 7. Similar patterns of spatial distribution were observed for Cu, Pb, and Hg. These elements showed a relatively high spatial variability. Furthermore, two different hot-spot areas were identified, i.e., in the north-eastern, and in the south-western to southern areas of the petrochemical complex. The elevated concentrations were likely created by the anthropogenic influences, particularly related to the wastes dumping into streams. As shown in Figure 3, two streams, dissecting the study area, are the main irrigation water sources for the arable land and contaminated by municipal sewage and industrial wastewater to a certain extent. In addition, the intensive application of metal-containing agrochemicals and organic composts derived from a variety of wastes and sludge during the long-term vegetable production also significantly contributed to the soil heavy metal pollution. However, the possibility of the atmospheric deposition of particles from the industrial emission, in some cases, could not be excluded. The prevailing winds originate from the north (in winter), and the pollutants likely deposited in these areas by the complex are released to the air.

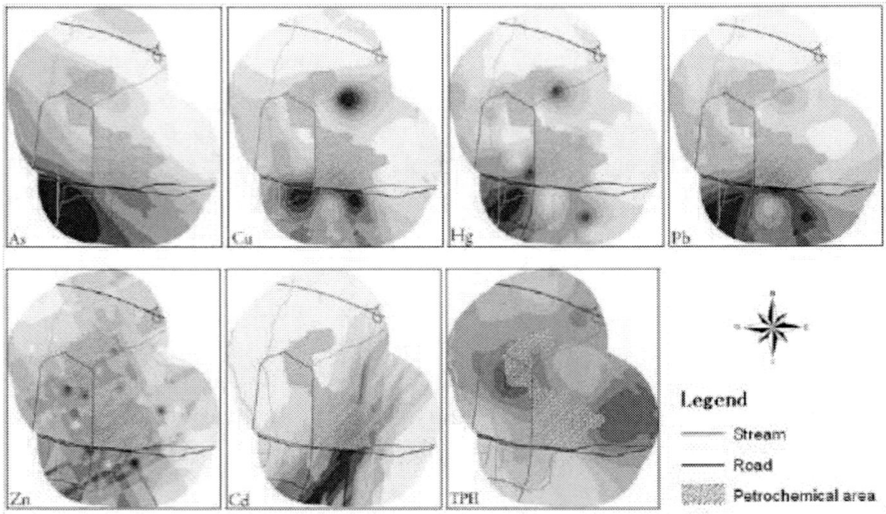

Figure 7. Spatial distribution of Zn, Cu, Pb, Cd, Hg, As, and TPH concentrations (mg kg^{-1}) in topsoils.

The spatial distribution pattern of Zn concentration was different from the metals mentioned above, showing a less variability. The hot-spot areas were located in the south and the area close to the petrochemical complex. The Zn concentration had a tendency to decrease with the increasing distance from the complex.

In case of Cd, the spatial distribution was very different from other metals. One hot-spot area located in the south of the complex was identified. The main utilization type of this area was vegetable field. Such anthropogenic inputs as intensive application of organic manures, phosphatic fertilizers, irrigation with metal-containing water were the dominant sources of the soil Cd pollution. Others also reported that numerous human activities, such as mining, smelting, waste disposal, wastewater irrigation, and phosphate fertilizer application had resulted in the release of significant quantities of Cd to the environment [12, 13, 17].

From the map of As spatial distribution, the highest concentration area was shown located in the south-western of the complex. However, the highest concentration was much less than the maximum allowable concentration of agricultural soil in China and did not pose a threat to the agricultural production.

According to the Environmental Quality Standard for Soils [42], the heavy metal concentration in agricultural soils should be lower than the maximum allowable concentration in order to guarantee agricultural production and protect human health, and mainly applicable to soils of farmland, vegetable field, orchard, etc. The maps of Cd, Cu, Hg, and Zn spatial distribution showed that the concentrations of these metals exceeded the maximum allowable concentration of agricultural soil in China in the hot-spot areas. It indicated that soil was polluted by heavy metals to some extent, affecting the safety of agricultural products. Therefore, some measures such as applying soil amendments to decrease the availability of these metals or planting crops with less accumulation of these metals should be taken to avoid or minimize human exposure to these toxic metals through food chains.

The spatial distribution map of TPH contamination in topsoils showed a relatively high spatial variability (Figure 7). As shown, the spatial distribution pattern for TPH was totally different, compared to metals Zn, Cu, Pb, Cd, Hg, and As. Two different hot-spot areas for TPH were identified, i.e., in the south-eastern and north-western areas of the petrochemical complex. The elevated concentrations might be due to the anthropogenic influences, particularly related to the accidental spills. There have been about 10 accidental explosions or burning accidents since the operation of the petrochemical complex in 1973, including the pipeline leakages in 2004 (July and August), the heavy oil spills in 2005 (August), and the oil-containing wastewater leakage caused by the explosion of sewage recycling cans in 2007 (January), which seriously contaminated the area. The air emission of petroleum might also be responsible for the hot-spot areas, even though the wind direction frequency (Figure 2) seemed not that much answer for the high concentration areas.

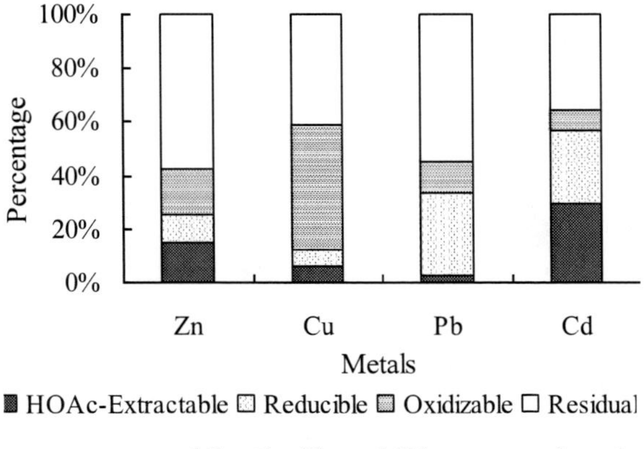

Figure 8. Average percentages of Zn, Cu, Pb, and Cd among various fractions in agricultural soils in the vicinity of a petrochemical complex in Guangzhou.

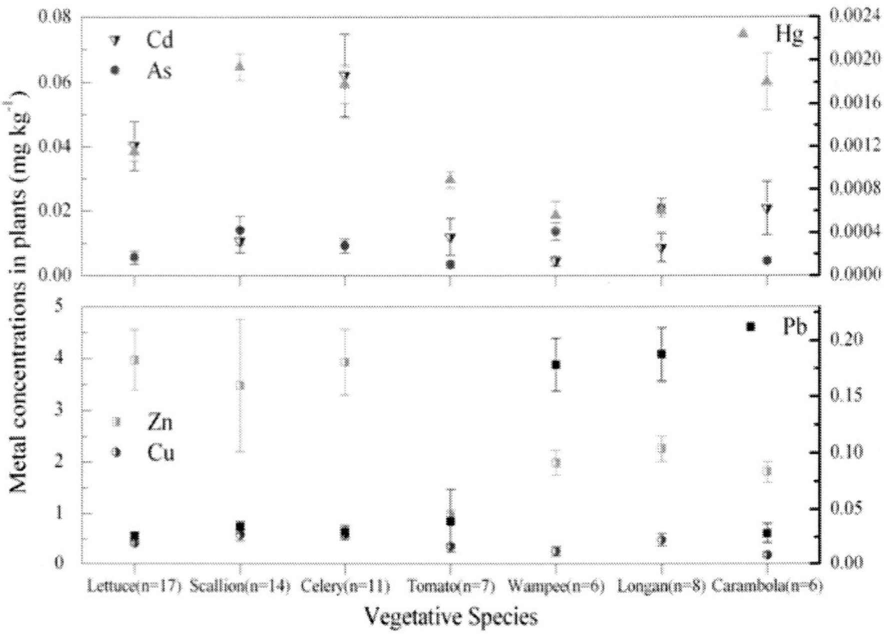

Figure 9. Metal concentrations in seven vegetative species (mg kg^{-1}).

3.5. Distribution of Metals among Different Fractions

The BCR extraction procedure was applied to assess the mobility and availability of metals (Zn, Cu, Pb, and Cd). The speciation results for those metals in 34 selected soil samples are shown in Figure 8. The recovery values for these four metals ranged from 87 to 124%, indicating a good agreement between the total concentration and the sum of the BCR results. The distribution of metals in four fractions varied greatly among the samples.

The distribution of various Cd fractions shows that the highest amount was found in the residual fraction (35.7% of total) and the lowest in the oxidizable fraction (7.7%). The HOAc extractable fraction (29.7%) was approximately equivalent to the reducible fraction (26.9%). In comparison, the greatest proportion of Cu was associated with the oxidizable fraction (46.2%), followed by the residual (41.4%). A very small proportion of Cu was found HOAc extractable (6.4%) and the reducible fraction made up the smallest proportion (6.0%). In case of Pb, it was primarily distributed in the residual fraction (54.5%), followed by the reducible (30.8%) and oxidizable fraction (12.1%), and the smallest amount of Pb was associated with the HOAc extractable fraction (2.6%). The dominating chemical form of Zn was associated with the residual fraction (57.9%). The residual fractions of metals were in the order of Zn > Pb > Cu > Cd.

3.6. Metal Concentrations in Plants

The concentrations of metals for various plants varied greatly among plant species (Figure 9). The metal concentrations were expressed on the basis of fresh weight (FW). The levels of Zn, Cu, Pb, Cd, Hg, and As in the edible parts of vegetables and fruits ranged from 0.45 to 19.36 (with the average of 3.01), from 0.033 to 1.89 (0.45), from 0.0026 to 0.31 (0.062), from 0.0014 to 0.15 (0.026), from 0.00015 to 0.0027 (0.0013), and from 0.0005 to 0.057 (0.010) mg kg^{-1}, respectively. The mean concentrations of Cu and Cd were the highest in celery and the mean concentrations of Pb and As were the highest in longan, while the highest concentrations of Zn and Hg were observed in lettuce and scallion, respectively. For Cd and Pb, 18.8% and 5.8% of all 69 samples exceeded the maximum permissible levels in food of China [56].

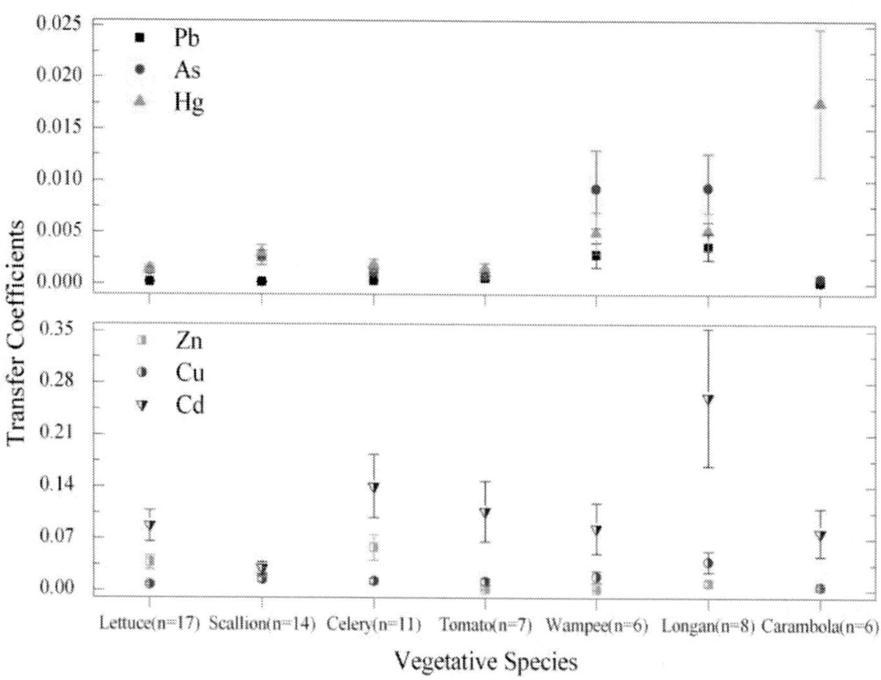

Figure 10. Transfer coefficients of metals in seven vegetative species.

3.7. Transfer Coefficients of Metals from Soil to Plants

The transfer coefficient (TC) not only quantifies the relative differences in the availability of metals to plants but also is a function of both soil and plant properties. The TC values for metals in the individual species of plant samples are shown in Figure 10. Results showed significant differences among metals and plant species. The highest TC values for Cu (0.033), Pb (0.0041), and As (0.0087) occurred in longan, for Zn (0.039) and Cd (0.18) in celery, and for Hg (0.019) in carambola, while the lowest TC values for Zn, As, and Hg were found in tomato. The overall TC values for metals were in the order of Cd > Zn > Cu > Hg > As > Pb.

3.8. Factors Affecting Metal Uptake by Plants

The PCA (Figure 11) was performed with the concentration of each metal in different fractions, the total concentration of each metal, the selected soil properties, and the concentration of each metal in plants in order to analyze the relationships among these indices. For Cd (Figure 11a), the first principal component, with the relative inertia of 54.6%, was positively determined by the Cd concentration in plants, the total concentration and four fractions Cd concentration, CEC, EC, and OM, respectively. The secondary principal component, with the relative inertia of 15.6%, was positively determined by clay content and negatively determined by pH. The Cd concentration in plants was shown close to the exchangeable fraction (HOAc extractable Cd) and the poorly labile fractions (oxidizable Cd, reducible Cd, and residual Cd), indicating significant correlations between them. Additionally, the plant Cd concentration correlated with OM, EC, and CEC. In comparison, Pb (Figure 11b) showed the inverse relationship between plant bioaccumulation and all four fractions of Pb even including the exchangeable fraction (HOAc extractable). For Zn (Figure 11c), the total and residual Zn fractions together with all the selected soil properties mainly explain the Zn concentration in plants, whereas other fractions were poorly related to the Zn accumulation by plants. In case of Cu (Figure 11d), there were no significant correlations between Cu concentrations in plants and any fractions. The concentration of Hg in soils was shown poorly related with the Hg accumulation by plants, while Hg in plants was related with soil pH (Figure 11e). The general trends for As (Figure 11f) were similar to Hg, even though the inverse relationship between As accumulation by plants and soil pH was observed.

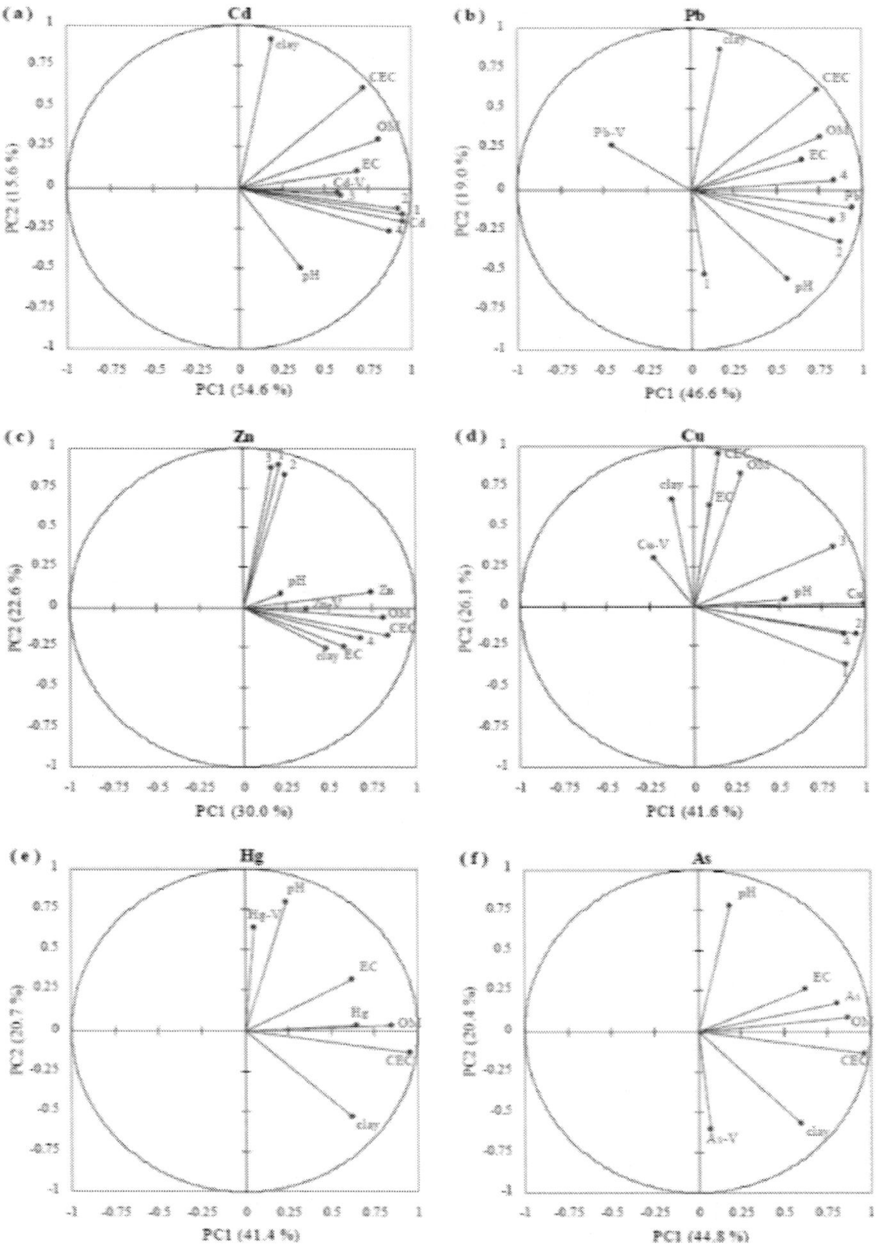

Figure 11. Plots of the first two principal components from the PCA of plant metal, soil metal, soil metal fractions, and selected soil parameters. (1, HOAc extractable metal; 2, reducible metal; 3, oxidizable metal; 4, residual metal; metal-V, metal concentration in plants).

In general, metals in the water/acid soluble and exchangeable fractions are considered readily and potentially mobile. The HOAc extractable fraction in the first step, which is approximately equivalent to the sum of water soluble, exchangeable, and carbonate-bound fractions, is the most active, mobile, and available phase of the metal [37, 38, 57]. In this study, in comparison to other metals, Cd was identified as being the most mobile element, occupying the highest percentage of HOAc extractable fraction, which is also in agreements with the previous studies [38, 57]. The average TC value was in the order of Cd > Zn > Cu > Hg > As > Pb, also indicating Cd in soil more available to plants, similar to previous studies [9, 33, 37, 58]. The oxidizable phase of metal is relatively stable under normal soil conditions, but it can be mobilized under strong oxidizing conditions due to the OM degradation, leading to the release of the soluble metal [37, 57]. Especially for Cu, a total of 46.2% soil Cu was associated with the oxidizable fraction, and Cu would become more mobile and available under strong oxidizing conditions. The reducible phase, similar to the oxidizable phase, can be mobilized under strong reduced conditions. On the other hand, metals in the residual fractions are entrapped within the crystal structure of minerals and thus represent the least mobile and available fraction [38, 57]. All four Pb-polluted plants were grown on soils where the Pb concentrations in topsoils did not exceed the maximum allowable concentrations in agricultural soil of China (data not shown). The main pathway of trace elements into plants is via their roots, but the atmospheric decomposition also influences the accumulation of metals [32, 59, 60]. In case of Cd-polluted plants, 9 out of 13 samples were grown on soils with the Cd concentrations exceeding the maximum allowable concentrations. The concentrations of other metals, however, did not exceed the safety limit in any of the tested vegetables and fruits, even though Hg, Cu, and Zn concentrations in topsoils of some areas exceeded the maximum allowable concentrations. The elevated levels of metals in soils may lead to their uptake by plants, which depends not only on metal contents in soils but also is determined by such other factors as soil pH, OM, clay content, CEC, and phosphate content, which cannot change the total amount of metals in soil but can significantly affect their bioavailability [37, 59]. Plant factors, such as plant species as well as growth period, also account for the uptake and translocation of metals [32, 59]. The concentrations of metals varied greatly among plant species (Figure 9). The levels of Cd and Pb in crops posed a potential risk to local residents via consumption, but not other metals. For 13 Cd-polluted plants, 5 and 6 plants were observed in lettuces and celeries, respectively, further suggesting lettuce and celery not cultivated on Cd-polluted soils.

In order to analyze the relationships between metal accumulation in plants and other indices, the PCA was performed. Cd is known for its easy uptake by plants, as shown by the positive significant correlations between plants, Cd concentrations, and all 4 fractions of Cd in soils. In case of Pb, the aerial deposition and foliar adsorption may account for the negative relationships between all 4 fractions of Pb in soil and Pb accumulation in plants. Metals could be accumulated in leaves much more efficiently than underground parts of vegetables [33]. In cases of Zn, Cu, Hg, and As, no significant positive relations were observed between the metal accumulation in plants and the HOAc extractable fraction of metals in soil, also suggesting the atmospheric deposition as another important way for the accumulation of metals in plants [32, 60]. Soil pH, on the other hand, is well known as a factor strongly influencing the metal mobility and availability through changing the equilibrium, partitioning and speciation of metals in the different soil phases [37, 61]. In general, the solubility, mobility, and potential plant availability of most metals increase with decreasing pH [61, 62]. In this study, the soil acidity significantly affected the contents of Pb and As in plants (Figs 11b and 11f), increasing the mobility and availability of those metals. In case of Hg (Figure 11e), however, the soil acidity decreased the Hg uptake. No correlation dependence between soil pH and concentrations of other metals in vegetables was observed. The OM content, CEC, and EC affected the accumulation of Zn, Cd, and Cu, while the clay content influenced the uptake of Cu, Hg, and As.

CONCLUSION

The concentrations of Zn, Cu, Pb, Cd, Hg, As, and TPH in agricultural soils around a petrochemical complex have been assessed. Significant differences of topsoil heavy metal levels between land-use types and heavy metal concentrations decreased with soil depth implied such human activities as wastewater irrigation and intensive application of sewage sludge and various agricultural chemicals as the main reason for heavy metal elevation in soils. Significant correlations between heavy metals in soils reflected that they had similar sources, compatible with the similar spatial distribution pattern. The indistinct gradients of decreasing concentrations of Cu, Pb, Cd, Hg, and As with distance from the complex as well as the distribution maps indicated little influence of the air emission of petroleum (except for Zn), further implying the anthropogenic input associated with the area. The spatial distribution of TPH along with the PCA showed that the accidental explosions or burning accidents mainly accounted for the TPH pollution.

Various industrial activities have resulted in the heavy metal accumulation in vegetables and fruits planted and cultivated around areas of the petrochemical complex in Guangzhou. The concentrations of Pb and Cd in a fraction of edible portions of investigated plants exceeded the maximum permissible levels in food of China. The BCR sequential extraction procedure and the PCA turned out the effective means for studying metal availability to plants. Further study is warranted to monitor heavy metal concentrations in aerosol and irrigation water in this area and their relationships with metal accumulation in plants.

This study also demonstrated the GIS mapping as a useful tool for evaluating the spatial variability of heavy metals in the affected soil, which would be of special significance for the remediation purposes since it may identify hotspot areas with high contaminant concentrations and help pinpoint exactly where the remediation measures should be taken.

Acknowledgment

This research was partly supported by the University of Macau Research Committee and the Macau Science and Technology Development Fund.

References

[1] Nadal, M., Mari, M., Schuhmacher, M., and Domingo, J.L. (2009). Multi-compartmental environmental surveillance of a petrochemical area: Levels of micropollutants. *Environment International*, *35*, 227-235.

[2] Nadal, M., Schuhmacher, M., and Domingo, J.L. (2007). Levels of metals, PCBs, PCNs and PAHs in soils of a highly industrialized chemical/petrochemical area: Temporal trend. *Chemosphere*, *66*, 267-276.

[3] Cetin, E., Odabasi, M., and Seyfioglu, R. (2003). Ambient volatile organic compound (VOC) concentrations around a petrochemical

complex and a petroleum refinery. *Science of the Total Environment, 312*, 103-112.

[4] Nadal, M., Schuhmacher, M., and Domingo, J.L. (2004). Levels of PAHs in soil and vegetation samples from Tarragona County, Spain. *Environmental Pollution, 132*, 1-11.

[5] Iturbe, R., Flores, C., Castro, A., and Torres, L.G. (2007). Sub-soil contamination due to oil spills in zones surrounding oil pipeline-pump stations and oil pipeline right-of-ways in Southwest-Mexico. *Environmental Monitoring and Assessment, 133*, 387-398.

[6] Zouir, A., Esteve-Turrillas, F.A., Chafik, T., Morales-Rubio, Á., and de la Guardia, M. (2009). Evaluation of the soil contamination of Tangier (Morocco) by the determination of BTEX, PCBs, and PAHs. *Soil and Sediment Contamination, 18*, 535-545.

[7] Stigter, J.B., de Haan, H.P., Guicherit, R., Dekkers, C.P., and Daane, M.L. (2000). Determination of cadmium, zinc, copper, chromium and arsenic in crude oil cargoes. *Environmental Pollution, 107*, 451-464.

[8] Wang, H.Y. and Stuanes, A.O. (2003). Heavy metal pollution in air-water-soil-plant system of Zhuzhou City, Hunan Province, China. *Water, Air and Soil Pollution, 147*, 79-107.

[9] Cui, Y.J., Zhu, Y.G., Zhai, R.H., Chen, D.Y., Huang, Y.Z., Qiu, Y., and Liang, J.Z. (2004). Transfer of metals from soil to vegetables in an area near a smelter in Nanning, China. *Environment International, 30*, 785-791.

[10] Liao, X.Y., Chen, T.B., Xie, H., and Liu, Y.R. (2005). Soil As contamination and its risk assessment in areas near the industrial districts of Chenzhou City, Southern China. *Environment International, 31*, 791-798.

[11] Bi, X., Feng, X., Yang, Y., Qiu, G., Li, G., Li, F., Liu, T., Fu, Z., and Jin, Z. (2006). Environmental contamination of heavy metals from zinc smelting areas in Hezhang County, western Guizhou, China. *Environment International, 32*, 883-890.

[12] Li, J.T., Qiu, J.W., Wang, X.W., Zhong, Y., Lan, C.Y., and Shu, W.S. (2006). Cadmium contamination in orchard soils and fruit trees and its potential health risk in Guangzhou, China. *Environmental Pollution, 143*, 159-165.

[13] Kisku, G.C., Barman, S.C., and Bhargava, S.K. (2000). Contamination of soil and plants with potentially toxic elements irrigated with mixed industrial effluent and its impact on the environment. *Water, Air and Soil Pollution, 120*, 121-137.

[14] Fytianos, K., Katsianis, G., Triantafyllou, P., and Zachariadis, G. (2001). Accumulation of heavy metals in vegetables grown in an industrial area in relation to soil. *Bulletin of Environmental Contamination and Toxicology*, *67*, 423-430.

[15] Chary, N.S., Kamala, C.T., and Raj, D.S. (2008). Assessing risk of heavy metals from consuming food grown on sewage irrigated soils and food chain transfer. *Ecotoxicology and Environmental Safty*, *69*, 513-524.

[16] Kachenko, A. and Singh, B. (2006). Heavy metals contamination in vegetables grown in urban and metal smelter contaminated sites in Australia. *Water, Air and Soil Pollution*, *169*, 101-123.

[17] Montagne, D., Cornu, S., Bourennane, H., Baize, D., Ratie, C., and King, D. (2007). Effect of agricultural practices on trace-element distribution in soil. *Communications in Soil Science and Plant Analysis*, *38*, 473-491.

[18] Martley, E., Gulson, B.L., and Pfeifer, H.R. (2004). Metal concentrations in soils around the copper smelter and surrounding industrial complex of Port Kembla, NSW, Australia. *Science of the Total Environment*, *325*, 113-127.

[19] Sichorova, K., Tlustos, P., Szakova, J., Korinek, K., and Balik, J. (2004). Horizontal and vertical variability of heavy metals in the soil of a polluted area. *Plant Soil and Environment*, *50*, 525-534.

[20] Schulin, R., Curchod, F., Mondeshka, M., Daskalova, A., and Keller, A. (2007). Heavy metal contamination along a soil transect in the vicinity of the iron smelter of Kremikovtzi (Bulgaria). *Geoderma*, *140*, 52-61.

[21] Onianwa, P.C. and Fakayode, S.O. (2000). Lead contamination of topsoil and vegetation in the vicinity of a battery factory in Nigeria. *Environmental Geochemistry and Health*, *22*, 211-218.

[22] Svendsen, M.L., Steinnes, E., and Blom, H.A. (2007). Vertical and horizontal distributions of Zn, Cd, Pb, Cu, and Hg in uncultivated soil in the vicinity of a zinc smelter at Odda, Norway. *Soil and Sediment Contamination*, *16*, 585-603.

[23] Krishna, A.K. and Govil, P.K. (2004). Heavy metal contamination of soil around Pali industrial area, Rajasthan, India. *Environmental Geology*, *47*, 38-44.

[24] Yang, C.Y., Cheng, B.H., Hsu, T.Y., Tsai, S.S., Hung, C.F., and Wu, T.N. (2000). Female lung cancer mortality and sex ratios at birth near a petroleum refinery plant. *Environmental Research*, *83*, 33-40.

[25] Yang, C.Y., Tsai, S.S., Cheng, B.H., Hsu, T.Y., and Wu, T.N. (2000). Sex ratio at birth associated with petrochemical air pollution in Taiwan. *Bulletin of Environmental Contamination and Toxicology*, *65*, 126-131.

[26] Lin, M.C., Yu, H.S., Tsai, S.S., Cheng, B.H., Hsu, T.Y., Wu, T.N., and Yang, C.Y. (2001). Adverse pregnancy outcome in a petrochemical polluted area in Taiwan. *Journal of Toxicology and Environmental Health - Part A*, *63*, 565-574.

[27] Yang, C.Y., Chang, C.C., Chuang, H.Y., Ho, C.K., Wu, T.N., and Chang, P.Y. (2004). Increased risk of preterm delivery among people living near the three oil refineries in Taiwan. *Environment International*, *30*, 337-342.

[28] Barona, A. and Romero, F. (1997). Relationships among metals in the solid phase of soils and in wild plants. *Water, Air and Soil Pollution*, *95*, 59-74.

[29] Li, J.H., Lu, Y., Yin, W., Gan, H.H., Zhang, C., Deng, X.L., and Lian, J. (2009). Distribution of heavy metals in agricultural soils near a petrochemical complex in Guangzhou, China. *Environmental Monitoring and Assessment*, *153*, 365-375.

[30] Turkdogan, M.K., Kilicel, F., Kara, K., Tuncer, I., and Uygan, I. (2003). Heavy metals in soil, vegetables and fruits in the endemic upper gastrointestinal cancer region of Turkey. *Environmental Toxicology and Pharmacology*, *13*, 175-179.

[31] Sharma, R.K., Agrawal, M., and Marshall, F.M. (2008). Heavy metal (Cu, Zn, Cd and Pb) contamination of vegetables in urban India: A case study in Varanasi. *Environmental Pollution*, *154*, 254-263.

[32] Voutsa, D., Grimanis, A., and Samara, C. (1996). Trace elements in vegetables grown in an industrial area in relation to soil and air particulate matter. *Environmental Pollution*, *94*, 325-335.

[33] Yan, S., Ling, Q.C., and Bao, Z.Y. (2007). Metals contamination in soils and vegetables in metal smelter contaminated sites in Huangshi, China. *Bulletin of Environmental Contamination and Toxicology*, *79*, 361-366.

[34] Nan, Z.R., Zhao, C.Y., Li, J.J., Chen, F.H., and Sun, W. (2002). Relations between soil properties and selected heavy metal concentrations in spring wheat (*Triticum aestivum* L.) grown in contaminated soils. *Water, Air and Soil Pollution*, *133*, 205-213.

[35] Tokalioglu, S. and Kartal, S. (2003). Relationship between vegetable metal and soil-extractable metal contents by the BCR sequential extraction procedure: Chemometrical interpretation of the data.

International Journal of Environmental Analytical Chemistry, *83*, 935-952.

[36] Mench, M., J. Vangronsveld, V. Didier, and H. Clijsters. (1994). Evaluation of Metal Mobility, Plant Availability and Immobilization by Chemical-Agents in a Limed-Silty Soil. *Environmental Pollution*, *86*, 279-286.

[37] Mocko, A. and Waclawek, W. (2004). Three-step extraction procedure for determination of heavy metals availability to vegetables. *Analytical and Bioanalytical Chemistry*, *380*, 813-817.

[38] Lu, Y., Zhu, F., Chen, J., Gan, H., and Guo, Y. (2007). Chemical fractionation of heavy metals in urban soils of Guangzhou, China. *Environmental Monitoring and Assessment*, *134*, 429-439.

[39] Institute of Soil Science, Chinese Academy of Sciences. (1978). Soil physical and chemical analyses. Shanghai, China: Shanghai Science and Technology (in Chinese).

[40] Li, J.H., Lu, Y., Shim, H., Deng, X.L,. Lian, J., Jia, Z.L., and Li, J.H. (2010). Use of the BCR sequential extraction procedure for the study of metal availability to plants. *Journal of Environmental Monitoring*, *12*, 466-471.

[41] Rauret, G., Lopez-Sanchez, J.F., Sahuquillo, A., Rubio, R., Davidson, C., Ure, A., and Quevauviller, P. (1999). Improvement of the BCR three step sequential extraction procedure prior to the certification of new sediment and soil reference materials. *Journal of Environmental Monitoring*, *1*, 57-61.

[42] National Environmental Protection Agency of China. (1995). Environment quality standard for soils (GB 15618-1995). Beijing: Chinese Standard.

[43] Richter, B.E. (2000). Extraction of hydrocarbon contamination from soils using accelerated solvent extraction. *Journal of Chromatography A*, *874*, 217-224.

[44] Dionex. (2004). Accelerated Solvent Extraction (ASE) of Hydrocarbon Contaminants (BTEX, Diesel, and TPH) in Soils.

[45] Li, J.H., Zhang, J.T., Lu, Y., Chen, Y.Q., Dong, S.S., and Shim, H. (2011). Determination of total petroleum hydrocarbons (TPH) in agricultural soils near a petrochemical complex in Guangzhou, China. *Environmental Monitoring and Assessment*, In press.

[46] GDPEMC (Guangdong Province Environmental Monitoring Center). (1990). Background values of soil of Guangdong Province. Guangzhou: Guangdong Province Environmental Monitoring Center (in Chinese).

[47] Adeniyi, A.A. and Afolabi, J.A. (2002). Determination of total petroleum hydrocarbons and heavy metals in soils within the vicinity of facilities handling refined petroleum products in Lagos metropolis. *Environment International*, *28*, 79-82.

[48] Iturbe, R., Flores, R.M., Flores, C.R., and Torres, L.G. (2004). TPH-contaminated Mexican refinery soil: Health risk assessment and the first year of changes. *Environmental Monitoring and Assessment*, *91*, 237-255.

[49] Soukup, D.A., Ulery, A.L., and Jones, S. (2007). Distribution of petroleum and aromatic hydrocarbons at a former crude oil and natural gas production facility. *Soil and Sediment Contamination*, *16*, 143-158.

[50] Adeniyi, A.A. and Owoade, O.J. (2010). Total petroleum hydrocarbons and trace heavy metals in roadside soils along the Lagos-Badagry expressway, Nigeria. *Environmental Monitoring and Assessment*, *167*, 625-630.

[51] Huang, S.W. and Jin, J.Y. (2008). Status of heavy metals in agricultural soils as affected by different patterns of land use. *Environmental Monitoring and Assessment*, *139*, 317-327.

[52] Yay, O.D., Alagha, O., and Tuncel, G. (2008). Multivariate statistics to investigate metal contamination in surface soil. *Journal of Environmental Management*, *86*, 581-594.

[53] Romic, M. and Romic, D. (2003). Heavy metals distribution in agricultural topsoils in urban area. *Environmental Geology*, *43*, 795-805.

[54] Burgos, P., Madejon, E., Perez-De-Mora, A., and Cabrera, F. (2008). Horizontal and vertical variability of soil properties in a trace element contaminated area. *International Journal of Applied Earth Observation and Geoinformation*, *10*, 11-25.

[55] Alloway, B.J. (1995). *Heavy metals in soils* (2nd ed.). London, Blackie Academic and Professional, Chapman and Hall.

[56] Ministry of Health of P.R. China. GB13106-91, 1991; GB4810-94, 1994; GB15201–94, 1994; GB15199-94, 1994; GB14935-94, 1994; GB2762-94, 1994 (in Chinese).

[57] Banat, K.M., Howari, F.M., and Al-Hamad, A.A. (2005). Heavy metals in urban soils of central Jordan: Should we worry about their environmental risks? *Environmental Research*, *97*, 258-273.

[58] Golia, E.E., Tsiropoulos, N.G., Dimirkou, A., and Mitsios, I. (2007). Distribution of heavy metals of agricultural soils of central Greece using the modified BCR sequential extraction method. *International Journal of Environmental Analytical Chemistry*, *87*, 1053-1063.

[59] Intawongse, M. and Dean, J.R. (2006). Uptake of heavy metals by vegetable plants grown on contaminated soil and their bioavailability in the human gastrointestinal tract. *Food Additives and Contaminants*, *23*, 36-48.
[60] Williams, P.N., Lei, M., Sun, G., Huang, Q., Lu, Y., Deacon, C., Meharg, A.A., and Zhu, Y.G. (2009). Occurrence and partitioning of cadmium, arsenic and lead in mine impacted paddy rice: Hunan, China. *Environmental Science and Technology*, *43*, 637-642.
[61] Blake, L. and Goulding, K.W.T. (2002). Effects of atmospheric deposition, soil pH and acidification on heavy metal contents in soils and vegetation of semi-natural ecosystems at Rothamsted Experimental Station, UK. *Plant and Soil*, *240*, 235-251.
[62] Harter, R.D. (1983). Effect of Soil-pH on Adsorption of Lead, Copper, Zinc, and Nickel. *Soil Science Society of America Journal*, *47*, 47-51.

In: Heavy Metal Sediments
Editor: Lucía H. Sanz
ISBN: 978-1-61470-690-8
© 2012 Nova Science Publishers, Inc.

Chapter 2

HEAVY METALS IN COASTAL SEA SEDIMENTS OF THE EASTERN ADRIATIC SEA

Jasmina Obhodas[1]* and Vladivoj Valkovic[2]
[1]Ruder Boskovic Institute, Zagreb, Croatia
[2]A.C.T.d.o.o., Zagreb, Croatia

ABSTRACT

Polluted sediments present a threat to the health of the marine ecosystem and indirectly to the public health. In this work, more than 700 samples of the surface sea sediments have been collected in the eastern Adriatic coastal sea and analyzed for number of elements in effort to screen the whole area and to find the patterns of pollution pressure. In order to manage the large number of data, the GIS based database containing chemical element concentrations and descriptive data such as type of location have been produced. Results presented as GIS concentration distribution maps clearly show the presence of "hot-spots", which are investigated in details. In addition, the background values for measured chemical elements have been evaluated as a scientific basis for assessment of enrichment factors and quality criteria standards. The approach to this problem that we have been developed for the eastern Adriatic Sea is proposed to be applied elsewhere in particular to the Mediterranean Sea region.

* Corresponding author: Dr. Jasmina Obhodas, Head of Laboratory for Nuclear Analytical Methods, Experimental Physics Department. Tel.: +385-1-4561-161, Fax.: +385-1-4680-239, Mob: +385-91-561-6269. e-mail: jobhodas@irb.hr.

INTRODUCTION

Adriatic is relatively narrow and shallow sea, but still one of the cleanest among seas of the Mediterranean, especially its eastern coast. Yet, the population growth and expanding of the coastal cities result in permanent devastation of maritime environment caused by intensifying industrialization, increase of maritime traffic, increase of waste and ballast waters, oil leakage events, intensive fishing, aquaculturing and tourism and occasionally by violation of the environmental legislative.

Pollution agents released in the environment can not be always immediately observed. But maritime environments, unlike most terrestrial, permanently accumulate pollutants in the sediments, therefore acute or long-term pollution can be easily detected by analyzing sediments with one of the various analytical methods available. In the work presented here, we used energy dispersive x-ray fluorescence (EDXRF) as a very suitable technique for large area screening. The aim was to obtain distributions of chemical elements in the coastal sea sediments not more than 10 m deep and to identify baseline values and eventual sources of pollution. EDXRF is a mature technique, very suitable for bulk samples. It is fast, nondestructive and requires small quantity of a sample, 2 g or less. EDXRF can be used for chemical concentration analysis for elements $11<Z<92$. The method has many other applications and has been often used in archeological and cultural heritage studies [1], health studies [2], planetary science missions [3] and in many other scientific fields [4].

Here we present results for the distribution of elements K, Ca, Ti, Cr, Mn, Fe, Ni, Cu, Zn, Ga, As, Br, Rb, Sr, Y, and Pb in sediments along the eastern Adriatic coast, especially in the most exposed regions such as marinas, shipyards, small harbors in the villages and ports. In addition, samples were collected in bays and beaches which were not consumed by human activities in order to distinguish natural concentrations from the anthropogenic contributions. The selection of analyzed elements was predetermined by the set-up of the used EDXRF instrumentation. Anyhow, the collected samples are archived at the Institute Ruđer Bošković, Zagreb, and can be used for additional analysis. In the future it is planned to supplement the geochemical characterization of the eastern Adriatic sediments with the distribution of elements Cd and Sn which will be analyzed by the Am excitation upgraded EDXRF technique and elements C, N, O, Al, Si, P, S, Hg, U and other will be analyzed by the neutron activation method.

The emphasis in discussion has been given to elements Cu, Zn, As and Pb which are used as biocides in antifouling paints intended for protection of underwater structures as the hulls of ships. Copper, zinc and lead are often classified in environmental studies as heavy metals with respect to their negative impact on the ecosystems and human health. In this way arsenic, which is actually the metalloid, is also very often discussed in literature as a heavy metal based on its toxicity and negative impacts on the health of the ecosystem [5-6]. In this research we found that these heavy metals are significant pollutants in the eastern Adriatic and many locations were ships are washed (the procedure that removes old paint layers from the boat hulls using the high pressure water) and painted are identified as pollution "hot spots". The results for Cu, Zn, As and Pb classified by categories defined by the type of the human activities will be discussed with respect to sediment quality and concentrations that are sustainable or not tolerable in marine sediments.

EXPERIMENTAL

Surface sediment samples have been collected from the year 2005 to 2008. For each sample the following data have been recorded: the exact GPS coordinates, location name, site description, depth and location category (port, beach, village, bay, marina service area, marina pier, other). At all locations the sediment sampling was performed by a grab sampler or by a diver, either from the land or from the boat. The analysis of the major and trace elements in the sediment samples was made on the fraction of 2 mm. After oven drying (105°C), sieving, grinding and homogenizing, they were sieved again (ϕ=45 μm) in order to press the tablets (F=9 t) weighting cca 2 g, with diameter of 2.5 cm. Concentrations of 16 elements (K, Ca, Ti, Cr, Mn, Fe, Ni, Cu, Zn, Ga, As, Br, Rb, Sr, Y, and Pb) were measured by the EDXRF with the W anode (60 kV, 60 mA) and the Mo secondary target in orthogonal geometry and measurement parameters of 35 kV and 35 mA. The irradiation time was 1000 s. X-ray spectra were collected with a Canberra Si(Li) detector (FWHM = 170 eV at 5.9 keV, Be window = 3 mil) and were analyzed by using the QXAS program package—direct comparison method [7]. The IAEA ''Lake Sediment'' was used as a standard reference material (SRM). The energies of the principal Kα emission lines were used for the quantization of all measured elements except for Pb, for which Pb Lβ was used. Still, the obtained results for As are ambiguous due to the spectral interference between As (Kα) and Pb

(Lα). The QA/QC procedures were done by using IAEA SRM SL-1. The SRM has been measured in regular intervals, after every ten samples. The relative measuring errors were K: 7.2 %, Ca: 8.1 %, Ti: 7.5 %, Cr: 8.6 %, Mn: 4.9, Fe: 2.5 %, Ni: 20.5 %, Cu: 20.5 %, Zn: 5.1 %, Ga: 23.3 %, As: 11.1 %, Br: 31.0 %, Rb: 9.8 %, Sr: 5.2 %, Y: 7.4 % and Pb: 20.5%. The errors were calculated based on SL-1 reference material and calibration used, vicinity of peaks determining elements of high concentrations and matrix effect.

The minimum detection limits (MDL) were: 1mg/kg for Ga, As, Rb, Y and Pb, 3 mg/kg for Cu, 5 for Cr, Ni and Br, 10 mg/kg for Fe and Zn, 13 mg/kg for Mn, 50 mg/kg for Sr, 80 mg/kg for Ti and 150 mg/kg for K and Ca. The MDL for different chemical elements varies with the energy of the characteristic X-rays and is proportional to the difference of the characteristic x-ray energy and the excitation energy. The MDLs were calculated from the standard reference material SL-1. The result is below the MDL if it is less than $3\sqrt{B}$, where B is the number of the background counts under the characteristic x-ray peak. The real MDL values depend on the absorption effects as well as on the overlap of spectral lines in each particular spectrum. The results between $3\sqrt{B}$ and $6\sqrt{B}$ are considered to have a high individual error. They have only an informative value.

The contour maps of Cu, Zn, As and Pb concentrations in sediments were produced by ESRI software “Arc View 9.3”, extension “Spatial Analyzer” – “Inverse Distance Weighted” method. Projection is Gauss-Kruger with central meridian 16.30’, linear proportion 0.9997 and y coordinate constant 250000. GPS coordinates of sampling locations were transformed in Gauss-Kruger coordinate system by using open source program “Field”. Distribution maps are showed in two views: 1. as gray graduated values ranged in 10 classes normally scaled and printed on the colored background contour maps with 6 classes logarithmically scaled. The 10th class of graduated values is printed in red in order to distinguish the most polluted locations and 2. as colored graduated values logarithmically scaled. These views were chosen in order to extract most important information. Particularly, the contour maps do not reflect the realistic situation in the environment due to the error embedded in the interpolation process. In the other hand, they clearly indicate anomalies in distribution like pollution sources and can point the direction of spreading trends. The graduated values give insight in sampling density and more accurately present the concentration distribution. Graduated values normally scaled will emphasize the highest concentrations while logarithmically scaled values give better insight in lower and moderately increased values.

For environmental studies logarithmic scaling is often more suitable since natural distributions are commonly shifted towards lower values with appearance of few outliers, unless if it is important to give an accent to anomalies observed. In addition, next to the maps there are frequencies histograms obtained by "Statistica 6.0" software, and graphical presentation of concentration variations according to 7 categories obtained by "SigmaPlot 8.0" software. 7 categories are: 1—bays (418 samples), 2—beaches (26 samples), 3—villages (67 samples), 4—ports (89 samples), 5—marinas-pier area (69 samples), 6—marina service areas (42 samples) and 7—others (sea mud, river inflows, 12 samples). The Category 1 samples were collected in uninhabited bays, usually accessible only by boat, at depths 0–10 m. The Category 2 samples were collected on public beaches, in most cases in shallow waters. The Category 3 samples were collected from the boat anchored in front of villages usually from the depths 5 to 10 m. The Category 4 samples were collected in the ports or harbors of settlements on the coast or on the islands, depth variable 5–10 m. The Category 5 samples were collected in marinas from one of the piers far from the service area, depth less than 8 m. The Category 6 samples were collected in marine service areas, usually around travel lift or boat washing areas, depths variable, but not deeper than 5 m. The Category 7 are different samples which do not fit any of the above criteria.

RESULTS

Descriptive statistics for all sediment locations are presented in Table 1. The Category 1 representing the unpolluted natural bays in the Eastern Adriatic Sea is presented in Table 2. These values can be used as baseline data. Results for individual chemical elements are described next in more detail.

Potassium

The range of concentrations measured in coastal sediments of the eastern Adriatic is 0,11 % - 3,4 %. Distribution is close to normal, showing no significant anomalies. Mean for total number of samples and categories is – total: 0,60 %, bays: 0,55 %, beaches: 0,38 %, settlements: 0,55 %, ports: 0,60 %, marina piers: 0,74 %, marina service areas: 0,69 %, peloids: 1,0 % and delta sediments: 1,3 %. The highest values are measured in delta of the Neretva river.

Table 1. Descriptive statistics for total number of samples

	Valid N	Mean	-95 % percentil	+95 % percentil	Geom. mean	Med.	Min.	Max.	St.D.	Skew.	Kurt.
K ppm	509	6022,2	5626,3	6418,1	4727,5	4535,0	1121	34120	4546,2	1,7	4,2
Ca %	722	22,2	21,5	22,9	19,3	24,5	0,72	39,00	9,2	-0,5	-0,9
Ti ppm	589	862,9	781,9	943,9	474,1	415,0	8,6	6283	1000,8	2,0	4,9
Cr ppm	435	57,3	52,2	62,3	42,4	44,8	6,1	418	53,4	3,4	16,5
Mn ppm	718	104,0	96,1	111,9	73,4	67,1	14,9	896,5	107,8	3,1	13,4
Fe %	723	0,6	0,6	0,7	0,3	0,3	0,008	5,55	0,7	2,3	7,0
Ni ppm	702	22,5	20,3	24,6	17,2	14,6	5,4	558	29,2	10,3	166,9
Cu ppm	723	241,4	124,1	358,7	27,1	17,1	1,8	31100	1606,4	13,1	211,6
Zn ppm	723	234,0	154,6	313,3	53,5	35,8	3,9	16800	1086,8	9,6	110,5
Ga ppm	202	10,4	8,5	12,3	7,7	6,6	3,1	117,3	13,5	5,0	29,7
As ppm	711	19,7	15,7	23,7	8,1	7,3	1,1	938	54,2	10,3	141,6
Br ppm	723	114,7	105,4	123,9	78,5	74,5	5,3	1667	126,4	4,6	38,4
Rb ppm	723	27,2	25,3	29,1	19,2	17,0	1,7	170,4	25,9	2,1	4,8
Sr ppm	723	1004,6	966,1	1043,0	840,9	987,0	112,2	3937	527,0	0,4	0,3
Y ppm	721	20,7	19,3	22,0	14,7	13,9	1,8	138,0	18,3	1,9	4,8
Pb ppm	721	41,8	31,8	51,8	13,8	9,9	2,1	2307	136,9	10,1	133,7

Table 2. Descriptive statistics for Category 1 – unpolluted bays

	Valid N	Mean	-95 % percentil	+95 % percentil	Geom. mean	Med.	Min.	Max.	St.D.	Skew.	Kurt.
K ppm	229	5514,9	4868,9	6160,8	4158,6	3453,0	1472	34120	4960,9	2,3	6,8
Ca %	417	25,4	24,7	26,2	23,5	27,9	1,150	39,00	7,6	-1,1	0,6
Ti ppm	298	614,3	514,0	714,7	322,9	237,5	8,6	6283	880,2	3,0	10,7
Cr ppm	193	57,8	49,6	66,0	44,4	45,0	7,1	418	57,8	4,4	23,3
Mn ppm	414	79,8	70,6	88,9	57,0	51,4	14,9	779	94,9	4,1	20,8
Fe %	418	0,3	0,3	0,4	0,2	0,2	0,008	5,55	0,6	4,3	26,0
Ni ppm	405	16,5	15,1	17,8	14,2	13,2	5,8	166	13,6	5,8	47,3
Cu ppm	418	16,4	15,5	17,2	15,0	14,6	1,8	104	9,1	5,0	34,7
Zn ppm	418	31,6	29,9	33,3	28,5	25,9	3,9	115,7	17,3	2,4	6,2
Ga ppm	58	6,7	5,7	7,7	6,1	5,8	3,1	25,6	3,7	3,0	11,8
As ppm	407	5,9	5,4	6,4	4,3	4,1	1,1	31,2	5,2	2,0	4,6
Br ppm	418	93,8	81,9	105,7	65,3	62,0	9,3	1667	123,8	6,8	69,8
Rb ppm	418	19,7	17,8	21,6	14,9	13,1	2,6	170,4	19,7	3,3	14,6
Sr ppm	418	1126,3	1079,0	1173,5	983,1	1180,0	112,2	2466	491,2	-0,1	-0,6
Y ppm	416	15,1	13,7	16,4	11,1	10,7	1,9	104,2	14,2	2,6	9,1
Pb ppm	416	9,0	8,2	9,7	7,1	6,3	2,1	65,6	8,0	3,2	13,9

In addition, locations with increased concentrations are settlement Seline situated in Velebit channel and marina Agana in Marina settlement. Increased values are found in peloids too. Concentrations between 0,2 % and 0,4 % are most frequent in the coastal sediments of the eastern Adriatic.

Calcium

Concentrations of calcium are in range 0,72 % to 39 %. Distribution is normal with no observable anomalies. Values decrease as anthropogenic input rise. Mean for total number of samples and according to categories is – total: 22,2 %, bays: 25,4 %, beaches: 23,6 %, settlements: 22,1 %, ports: 17,0 %, marina piers: 16,6 %, marina service areas: 12,6 %, peloids: 11,0 % and delta sediments: 14,1 %. The lowest values are measured in San Marino beach on the island of Rab and highest in Žalić port on the island of Silba. Areas with lower concentrations are island of Jabuka which has volcanic origin, island of Susak which has alluvial origin, Kvarner bay, Seline bay, island of Mljet (southern part) and island of Koločep. Lower values of calcium are found in peloids. Concentrations between 20 % and 30 % are most frequent in the coastal sediments of the eastern Adriatic.

Titanium

The range of measured concentrations is between 8,6 mg/kg and 6283 mg/kg. Distribution of titanium is slightly shifted from normal. Increase in concentrations caused by anthropogenic contribution due to the pigments in antifouling paints is notable. Mean for total number of samples and categories is – total: 862,9 mg/kg, bays: 614,3 mg/kg, beaches: 619,3 mg/kg, settlements: 787,7 mg/kg, ports: 995,4 mg/kg, marina piers: 1276,8 mg/kg, marina service areas: 1628,7 mg/kg, peloids: 2124,0 mg/kg and delta sediments: 1888,2 mg/kg. The highest concentrations were found in Zagrob bay on the eastern coast of Istria peninsula and in Limski kanal on the western coast of Istria peninsula. Besides in marinas, higher concentrations are found in Seline settlement and on the island of Jabuka. Concentrations between 50 mg/kg and 5000 mg/kg are most frequent in the coastal sediments of the eastern Adriatic.

Chromium

The range of measured concentrations is between 6,1 mg/kg and 418 mg/kg. Distribution is shifted from normal, but it is not clear if the shift can be assigned to the anthropogenic contribution. Anomalies could refer to geological background, even though, the mean value for settlements is slightly higher when compared to other categories. Mean for total number of samples and categories is – total: 57,3 mg/kg, bays: 57,8 mg/kg, beaches: 44,7 mg/kg, settlements: 70,0 mg/kg, ports: 52,8 mg/kg, marina piers: 47,8 mg/kg, marina service areas: 62,3 mg/kg, peloids: 90,3 mg/kg and delta sediments: 57,3 mg/kg. The highest concentrations were measured on the island of Koločep. The area of Elafiti islands, the bay of Ston and the southern part of the island of Mljet make the region distinguished by the highest concentrations of Cr in the eastern Adriatic (>100 mg/kg). The Kvarner bay, the island of Brač and northern part of the island of Hvar have the lowest Cr concentrations (<25 mg/kg). Concentrations between 20 and 60 mg/kg are most frequent in the coastal sediments of the eastern Adriatic.

Manganese

Concentrations of manganese are in range 14,9 mg/kg to 896,5 mg/kg. Distribution is shifted from normal and anomalies caused by anthropogenic contributions are notable. Mean for total number of samples and categories is – total: 104,0 mg/kg, bays: 79,8 mg/kg, beaches: 94,8 mg/kg, settlements: 118,8 mg/kg, ports: 130,0 mg/kg, marina piers: 117,1 mg/kg, marina service areas: 215,8 mg/kg, peloids: 236,1 and delta sediments: 292,8 mg/kg. The highest values are measured on the island of Mljet in the ACI marina Jezera – service area. Increased concentrations are found on the island of Jabuka, in the delta of Neretva river, in Ploče harbor, in towns Split, Šibenik and Rijeka, on the island of Susak and in the Novi Vinodolski settlement. The most frequent concentrations are in range from 10 to 400 mg/kg.

Iron

The range of measured concentrations is between 0,008 % and 5,55 %. Distribution of iron is bimodal. Mean for total number of samples and categories is – total: 0,6 %, bays: 0,3 %, beaches: 0,5 %, settlements: 0,8 %, ports: 0,9 %, marina piers: 1,0 %, marina service areas: 1,6 %, peloids: 1,4 % and delta sediments: 1,1 %. The lowest values are measured at the island of Dugi otok and highest at the island of Jabuka. Increased concentrations were measured in all coastal towns like Pula, Rijeka, Zadar, Šibenik, Split, Ploče and Dubrovnik. Higher concentrations were found in marinas' service areas, in peloids, in sediments of the Neretva delta, the island of Susak and in Seline settlement. The most frequent concentrations present in the sediments of the eastern Adriatic are between 0,07 to 0,25 % and between 0,75 to 2,5%.

Nickel

The range of measured concentrations is between 5,4 mg/kg and 558 mg/kg. Distribution is heavily shifted from normal with distinguished anthropogenic contribution. Mean for total number of samples and categories is – total: 22,5 mg/kg, bays: 16,3 mg/kg, beaches: 18,4 mg/kg, settlements: 23,9 mg/kg, ports: 23,8 mg/kg, marina piers: 25,3 mg/kg, marina service areas: 71,7 mg/kg, peloids: 36,6 mg/kg and delta sediments: 24,5 mg/kg. The highest values are measured in ACI marina Vodice – service area. Increased concentrations are present in the coastal towns Pula, Rijeka, Zadar, Šibenik, Split, Ploče and Dubrovnik and in marinas' service areas in general. Slightly increased concentration can be found in peloids and on the island of Jabuka, Seline settlement and in Dubrovnik coastal sea area (southern part of the island of Mljet, the Ston bay, Elafiti islands). The most frequent concentrations present in the sediments of the eastern Adriatic are between 9 and 20 mg/kg.

Copper

The range of measured concentrations is between 1,8 mg/kg and 31100 mg/kg. Distribution is heavily shifted from normal with distinguished anthropogenic contribution. Mean for total number of samples and categories is – total: 241,4 mg/kg, bays: 16,4 mg/kg, beaches: 19,5 mg/kg, settlements: 71,2 mg/kg, ports: 121,7 mg/kg, marina piers: 90,4 mg/kg, marina service areas: 3448,9 mg/kg, peloids: 26,5 mg/kg and delta sediments: 61,2 mg/kg. The lowest values were found on the island of Lastovo, and the highest in marina ACI Vodice in service area. In general, the increased concentrations were measured in marinas' service areas. Slightly increased concentrations were measured on the islands of Jabuka and Sušac. The most frequent concentrations present in the sediments of the eastern Adriatic are between 5 and 50 mg/kg.

Zinc

The range of measured concentrations is between 3,9 mg/kg and 16800 mg/kg. Distribution is heavily shifted from normal with distinguished anthropogenic contribution. Mean for total number of samples and according to categories is – total: 234,0 mg/kg, bays: 31,6 mg/kg, beaches: 41,1 mg/kg, settlements: 161,3 mg/kg, ports: 205,9 mg/kg, marina piers: 175,2 mg/kg, marina service areas: 2687,5 mg/kg, peloids: 63,9 mg/kg and delta sediments: 66,4 mg/kg. The lowest values were measured in the Javorina bay and in Velebit channel. The highest values were measured in marina ACI Vodice in service area. In general, the increased concentrations were measured in marinas' service areas. Slightly increased concentrations were measured on the island of Jabuka. The most frequent concentrations present in the sediments of the eastern Adriatic are between 5 and 50 mg/kg.

Gallium

The range of measured concentrations is between 3,1 mg/kg and 117,3 mg/kg. Distribution is shifted from normal caused by the notable anthropogenic contribution in marinas' service areas. Mean for total number of samples and categories is – total: 10,4 mg/kg, bays: 6,7 mg/kg, beaches: 7,4 mg/kg, settlements: 13,8 mg/kg, ports: 8,0 mg/kg, marina piers: 8,3 mg/kg, marina service areas: 26,9 mg/kg, peloids: 9,0 mg/kg and delta sediments: 5,6 mg/kg. The highest values were measured in marina «Yacht Club Opatija», in service area. In general, the increased concentrations were measured in marinas' service areas. Slightly increased concentrations were measured on the island of Jabuka. The most frequent concentrations present in the sediments of the eastern Adriatic are between 4 and 6 mg/kg.

Arsenic

The range of measured concentrations is between 1,1 mg/kg and 938 mg/kg. Distribution is heavily shifted from normal with distinguished anthropogenic contribution, especially in marinas' service areas. Mean for total number of samples and categories is – total: 19,7 mg/kg, bays: 5,9 mg/kg, beaches: 7,2 mg/kg, settlements: 45,1 mg/kg, ports: 25,9 mg/kg, marina piers: 25,5 mg/kg, marina service areas: 101,8 mg/kg, peloids: 12,5 mg/kg and delta sediments: 9,0 mg/kg. The highest values were measured in the town of Pula. In general, the increased concentrations were measured in marinas' service areas. Slightly increased concentrations were measured on the island of Jabuka. The most frequent concentrations present in the sediments of the eastern Adriatic are between 1 to 50 mg/kg.

Bromine

The range of measured concentrations is between 5,3 mg/kg and 1667 mg/kg. Distribution is shifted from normal with notable anthropogenic contribution. Mean for total number of samples and a categories is – total: 114,7 mg/kg, bays: 93,8 mg/kg, beaches: 44,8 mg/kg, settlements: 97,6 mg/kg, ports: 140,7 mg/kg, marina piers: 193,0 mg/kg, marina service areas: 202,4 mg/kg, peloids: 167,4 mg/kg and delta sediments: 97,0 mg/kg. The highest values were measured on the Oštra cape and lowest in the Drvarica bay in Velebit channel. Increased concentrations of bromine were measured in peloids and in marinas. The most frequent concentrations present in the sediments of the eastern Adriatic are between 75 to 250 mg/kg.

Rubidium

The measured concentrations are between 1,7 mg/kg and 170,4 mg/kg. Distribution is very close to normal, although the anthropogenic contribution is present. Mean for total number of samples and categories is – total: 27,2 mg/kg, bays: 19,7 mg/kg, beaches: 14,2 mg/kg, settlements: 27,2 mg/kg, ports: 34,3 mg/kg, marina piers: 50,3 mg/kg, marina service areas: 46,8 mg/kg, peloids: 70,8 mg/kg and delta sediments: 51,1 mg/kg. The lowest values were measured in Žurkovo settlement and highest on the island of Veliki Brijun. Slightly increased concentrations can be found in peloids, marinas, on the island of Susak, Seline settlement and on the island of Jabuka. The most frequent concentrations present in the sediments of the eastern Adriatic are between 9 and 20 mg/kg.

Strontium

The range of measured concentrations is between 112,2 mg/kg and 3937 mg/kg. Distribution is normal, hence in environments consumed by the anthropogenic contribution its concentrations tend to decrease. Mean for total number of samples and categories is – total: 1004,6 mg/kg, bays: 1126,3 mg/kg, beaches: 682,0 mg/kg, settlements: 1020,5 mg/kg, ports: 695,7 mg/kg, marina piers: 965,4 mg/kg, marina service areas: 737,2 mg/kg, peloids: 676,8 mg/kg and delta sediments: 10038,2 mg/kg. The lowest values were measured on the island of Cres in the Homutići bay and highest on the island of Dugi otok in the Telašćica Nature Park, location Slano lake. Slightly increased concentrations were measured in sediments of Neretva delta. The lowest values were found in the area of the Kvarner bay. This reflects the distribution of the mud-send fraction which ratio is > 1:1 [8]. Lower concentrations were measured on the islands of Jabuka, Vis, Sušac, Palagruža, Brač and in the middle and the sothern part of the island of Mljet. The most frequent concentrations present in the sediments of the eastern Adriatic are between 1000 and 3000 mg/kg.

Yttrium

The range of measured concentrations is between 1,8 mg/kg and 138 mg/kg. Distribution is very close to normal, although the anthropogenic contribution is present. Mean for total number of samples and categories is – total: 20,7 mg/kg, bays: 15,1 mg/kg, beaches: 17,6 mg/kg, settlements: 21,1 mg/kg, ports: 25,7 mg/kg, marina piers: 37,8 mg/kg, marina service areas: 28,3 mg/kg, peloids: 51,9 mg/kg and delta sediments: 60,4 mg/kg. The highest values were measured in the delta of the river Neretva. In general, slightly increased concentrations can be found in peloids, marinas, on the Susak island, Seline settlement and on the island of Jabuka. The most frequent concentrations present in the sediments of the eastern Adriatic are between 9 and 20 mg/kg.

Lead

The range of measured concentrations is between 2,1 mg/kg and 2307 mg/kg. Distribution is heavily shifted from normal with distinguished anthropogenic contribution. Mean for total number of samples and according to categories is – total: 41,8 mg/kg, bays: 9,0 mg/kg, beaches: 11,1 mg/kg, settlements: 100,3 mg/kg, ports: 52,7 mg/kg, marina piers: 52,8 mg/kg, marina service areas: 258,3 mg/kg, peloids: 21,1 mg/kg and delta sediments: 15,4 mg/kg. The highest values were measured in the town of Pula. In general, the increased concentrations were measured in marinas' service areas. The most frequent concentrations present in the sediments of the eastern Adriatic are between 7 and 25 mg/kg.

DISCUSSION

Elements Cu, Zn, As and Pb are recognized as the most harmful heavy metal pollutants in the eastern Adriatic coast [7-9]. Their distribution in the coastal sediments is strongly affected by the usage of antifouling paints for the protection of ship hulls against fouling. Other anthropogenic sources of these elements are sewage outfalls and agricultural practices. For the long time Cu, Zn, As and Pb in antifouling paints, herbicides and fungicides were considered to be harmful only for lower organisms. Unfortunately, we have learned that amount of bioaccumulation in lower organisms which enter into the humans' food chain can be that high to cause acute or chronic intoxication in humans. Figures 1 - 4 show concentration distribution maps indicating sources of sediment contamination by toxic elements Cu, Zn, As and Pb.

Table 3 shows descriptive statistics of these elements for different categories distinguished by the type of the environment.

The results show that the contents of Zn, Cu, As and Pb in sediments close to the service areas of marinas or ship yard services are several orders of magnitude greater than the background values presented in Table 2. These hot-spots are easily distinguished in the concentration distribution maps for Zn, Cu, As and Pb, as it is presented in Figures 1 - 4. The obtained distribution maps are not representative regarding the range of spreading, but they can distinctly visualize the degree of variation. Actually, the high concentrations in measured sources of pollution rapidly decrease along the transport pathways of suspended matter.

Table 3. Descriptive statistics of elements Cu, Zn, As and Pb for different categories

Element (mg/kg)	Category	Valid N	Mean	-95 % percentil	+95 % percentil	Geom. mean	Med.	Min.	Max.	St.D.	Skew.	Kurt.
Cu	Beaches	26	19,5	14,9	24,2	17,5	16,9	6,5	63,7	11,6	2,8	8,8
	Settlements	67	71,2	24,6	117,8	33,7	23,9	4,9	1544	190,9	7,2	55,6
	Ports	89	121,7	42,8	200,5	43,8	35,6	4,8	3044,0	374,5	6,4	45,2
	Marina piers	69	90,4	57,0	123,8	57,0	54,3	7,5	1066	139,0	5,4	36,2
	M. service areas	42	3448,9	1636,3	5261,6	964,0	1393,5	20,6	31100	5816,8	3,1	12,1
	Other-Pelloids	7	26,5	17,0	36,0	24,7	26,2	12,9	41,7	10,3	0,2	-1,2
	Other-D.Sediments	5	61,2	-52,1	174,5	31,4	19,2	13,8	223,8	91,3	2,2	4,8
Zn	Beaches	26	41,1	31,0	51,2	35,6	33,0	12,7	115	25,0	1,7	2,4
	Settlements	67	161,3	55,2	267,4	78,5	63,3	22,7	3487	435,1	7,1	53,7
	Ports	89	205,9	128,3	283,5	112,3	103,5	11,7	2840,0	368,3	5,3	33,4
	Marina piers	69	175,2	133,9	216,5	126,5	104,5	27,7	804	172,0	2,1	4,2
	M. service areas	42	2687,5	1541,7	3833,3	977,9	1142,5	43,3	16800	3677,0	2,1	4,7
	Other-Pelloids	7	63,9	43,0	84,7	60,0	62,7	27,0	102,9	22,5	0,2	2,2
	Other-D. Sediments	5	66,4	29,8	103,0	59,0	77,6	21,8	91,9	29,5	-1,0	-0,3
As	Beaches	26	7,2	4,7	9,6	5,4	6,0	1,6	29,4	6,1	2,2	6,3
	Settlements	66	45,1	13,2	76,9	16,9	13,1	1,8	938	129,6	6,0	38,0
	Ports	89	25,6	20,7	30,5	17,6	20,1	2,1	136,5	23,2	2,2	6,7
	Marina piers	69	25,5	19,5	31,4	18,7	18,0	2,9	132,7	24,8	2,6	7,4
	M. service areas	42	101,8	68,2	135,4	63,3	74,8	9,1	524	107,9	2,3	6,0

Table 3. (Continued)

Element (mg/kg)	Category	Valid N	Mean	-95 % percentil	+95 % percentil	Geom. mean	Med.	Min.	Max.	St.D.	Skew.	Kurt.
	Other-Pelloids	7	12,5	8,9	16,2	12,0	12,1	6,5	19,5	4,0	0,4	1,6
	Other-D. Sediments	5	9,0	1,4	16,6	7,5	8,6	3,4	19,0	6,1	1,4	2,2
Pb	Beaches	26	11,1	6,1	16,1	7,7	6,6	3,0	44,1	12,4	2,0	2,6
	Settlements	67	100,3	23,5	177,1	30,5	25,8	2,5	2307	315,0	6,1	39,5
	Ports	89	52,7	39,7	65,7	31,5	34,3	3,7	396	61,8	3,0	12,0
	Marina piers	69	52,8	38,8	66,8	36,4	33,6	5,4	308,5	58,3	2,8	7,9
	M. service areas	42	258,3	165,4	351,3	146,2	169,9	15,2	1430,	298,3	2,4	6,4
	Other-Pelloids	7	21,1	15,6	26,7	20,5	18,6	15,4	30,2	6,0	1,1	-0,9
	Other-D. Sediments	5	15,4	0,7	30,0	11,4	15,2	2,7	33,9	11,8	1,0	1,4

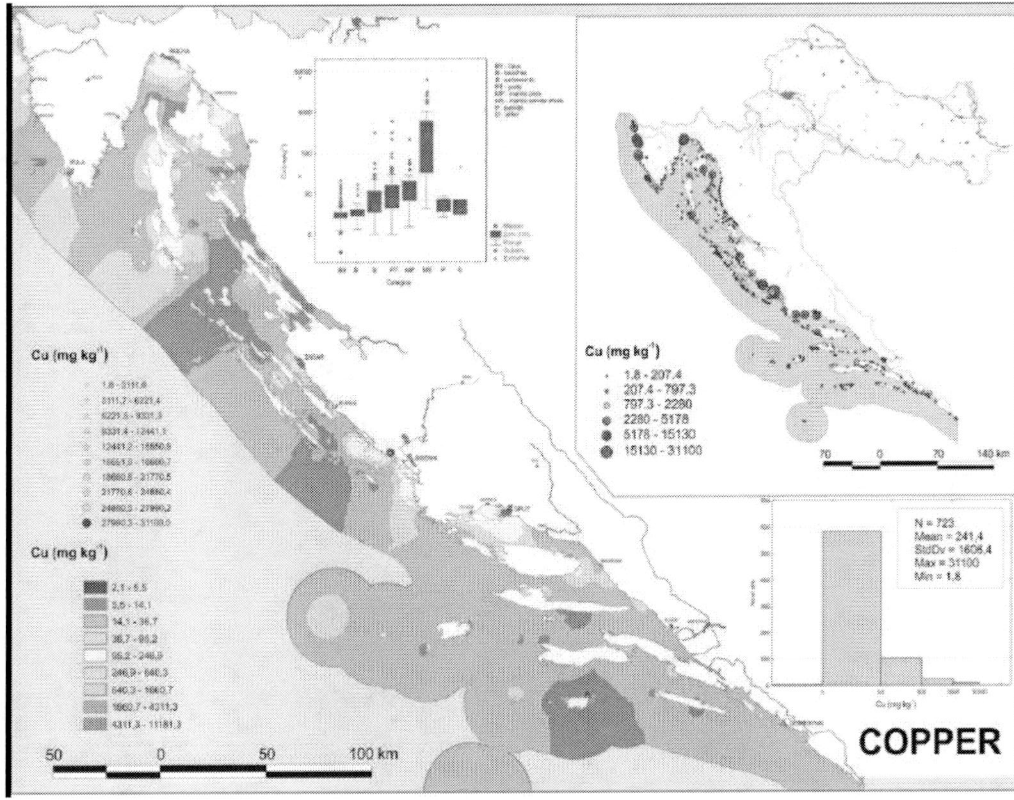

Figure 1. Distribution of Cu in the eastern Adriatic Sea.

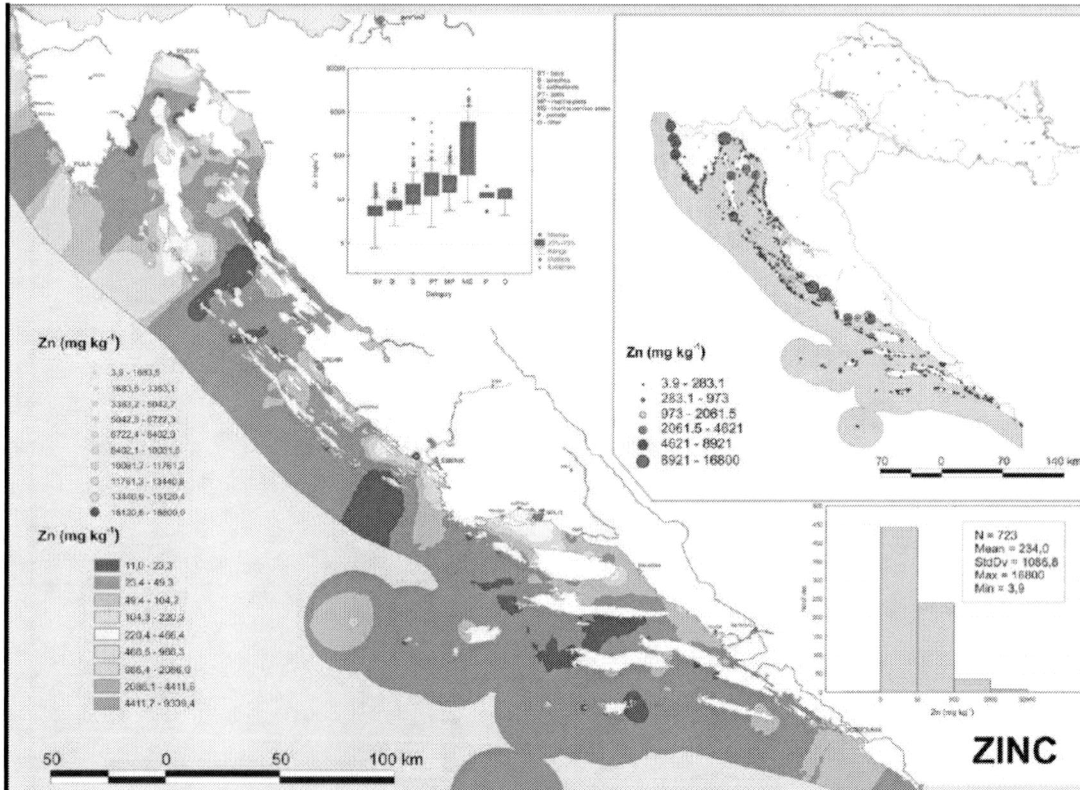

Figure 2. Distribution of Zn in the eastern Adriatic Sea.

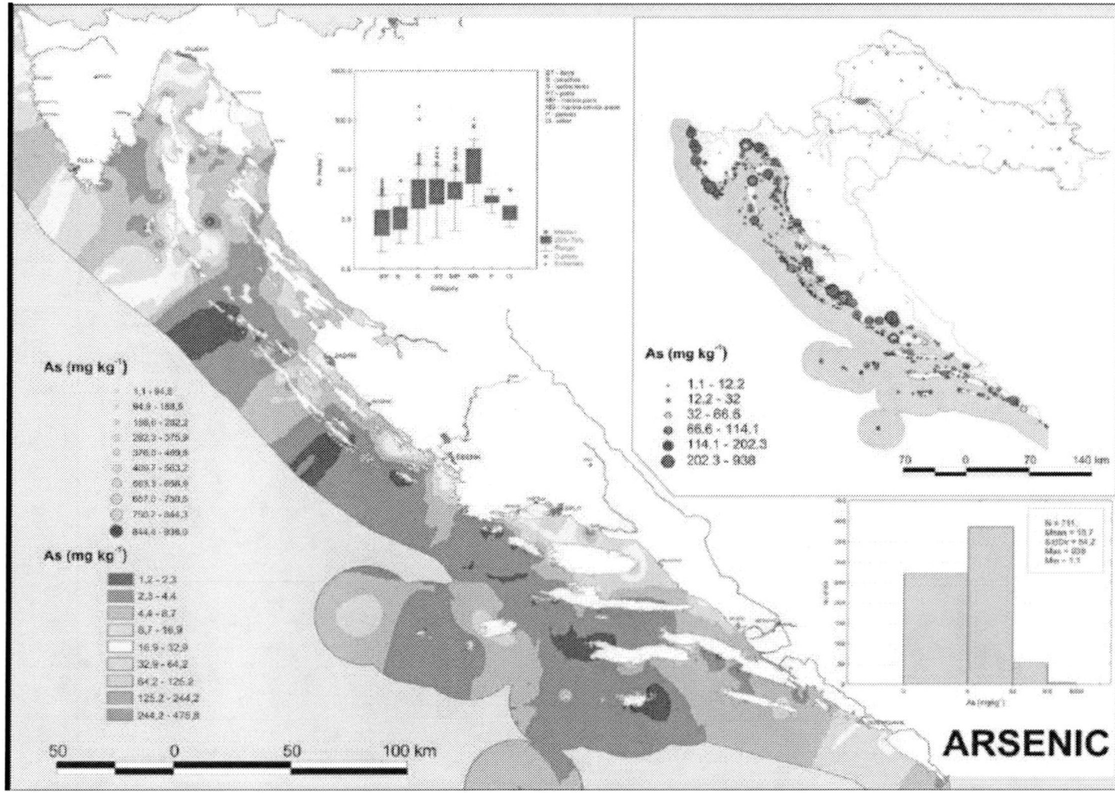

Figure 3. Distribution of As in the eastern Adriatic Sea.

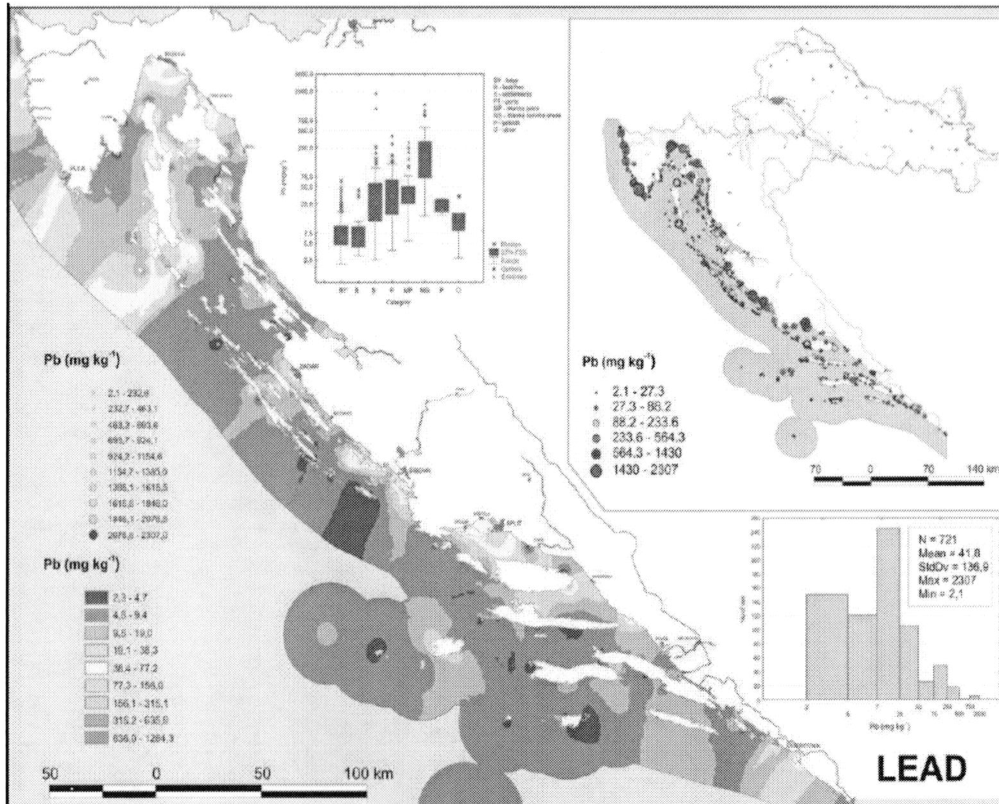

Figure 4. Distribution of Pb in the eastern Adriatic Sea.

The transport depends on the maritime conditions such are sea currents, sedimentation rate and sea depth. Concentrations usually drop several orders of magnitude just 100 m from the pollution source. This is by no means the limit of marinas' negative environmental impact. It has been proven that very small amounts of biocides concentrations in the sea water can have an impact on marine organisms. According to Myers et al. [9], the concentration of 1 mg/l is toxic to marine macro alga *Hormosira banksii.* In the vicinity of boat berths and in marinas, the Cu concentration in sea water can reach up to 20 mg/l [10]. Considering this, planning marinas in the vicinity of aquacultures and recreational beaches should be taken with most care.

In order to propose protective values of Cu, Zn, As and Pb in coastal sediments for Croatia, we present a quick overview of sediment quality criteria in EU and world. Sediment quality criteria guidelines (SQCG) adopted worldwide highly vary in multiple aspects. SQCG are derived based on sediment chemistry (distribution of contaminant concentrations, equilibrium partitioning, mathematical modeling), environmental effects (bioassay, community matrices, human exposure, toxicological modeling) or combination thereof. Often they are adopted based on other countries guidelines and adjusted for specific contaminants. Diversity of terms (screening values, guidance values, target and intervention values, maximum acceptable concentrations, cut off values, trigger values, environmental quality objectives, etc.) reflects the lack of a coherent framework and points that SQCGs are derived for different purposes such as classification of contaminants, health risk assessment, action levels for dredged material, clean-up standards etc.

SQCGs are best defined in the frame of dredging activities. Dredging in EU countries is regulated by the „Water Framework Directive" known as WFD or 2000/60/EC Directive. WFD demands that before disposal dragged material has to be analyzed and classified according to the critical parameters concentrations values. Most countries recognize two limiting values:

- Action Level 1 or lower limit - concentrations allowed for the sea disposal. Concentrations below this limit are considered harmless.
- Action Level 2 or upper limit - concentrations exceeding this limit are not allowed for disposal into sea.

Concentrations falling between the lower and the upper limit may trigger further investigation of the material proposed for disposal into the sea. National Action Levels are defined for the north-east Atlantic (EU regulative) and for Pacific (USA, Australia and New Zealand, Canada, Hong Kong).

Tables 4 and 5 present range of national Action Levels 1 and 2 for Cu, Zn, As and Pb defined for the north-east Atlantic and for Pacific respectively.

Table 4. Range of EU national Action Levels 1 and 2 for Cu, Zn, As and Pb concentrations in dredged materials (north – east Atlantic)

Element (mg/kg)	Action Level 1	Action Level 2
As	20 - 80	29 - 1000
Cu	20 - 150	60 - 1500
Pb	50 - 120	110 - 1500
Zn	130 - 700	365 - 10000

Table 5. Range for USA, Australian and N. Zealand, Canada and Hong Kong national Action Levels 1 and 2 for Cu, Zn, As and Pb concentrations in dredged materials (Pacific)

Element(mg/kg)	Action Level 1	Action Level 2
As	6 - 57	33 - 70
Cu	16 - 390	108 - 270
Pb	30,2 - 450	112 - 250
Zn	120 - 410	271 - 820

Table 6. Dredged material classification and disposal procedures for Portugal [11]

Elements mg/kg	Class 1	Class 2	Class 3	Class 4	Class 5
Copper (Cu)	<35	[35-150]	[150-300]	[300-500]	>500
Zink (Zn)	<100	[100-600]	[600-1500]	[1500-5000]	>5000
Arsenic (As)	<20	[20-50]	[50-100]	[100-500]	>500
Lead (Pb)	<50	[50-150]	[150-500]	[500-1000]	>1000
Description	Clean material	Contamination in traces	Slightly contaminated	Contaminated	Very contaminated
Disposal	Aquatic (including beaches)	Aquatic	Aquatic (obligating monitoring)	Land disposal (obligating monitoring)	Waste disposal site (special treatment).

For the eastern Adriatic we propose definition of background values (Action Level 1) by adopting mean values calculated for unpolluted bays (Table 2) and magnified for 3 standard deviations. In such way background values obtained for Cu, Zn, As and Pb are very similar to background values defined for Portugal. In Portuguese dredged material classification presented in Table 6, sediments are classified into 5 classes. Each class has its own color which can be used for highlighting values in graphical and table presentations. This classification corrected for concentrations of Cu and As in Class 1 to the values of <45 mg/kg and <25 mg/kg, respectively, could be adopted for Croatia SQCGs, although background values in the eastern Adriatic for Zn and Pb are slightly lower compared to Portuguese classification.

Concentration values for all four elements have the same trend, increasing as we move from bays and beaches to settlements, ports, marina-piers and marina-service areas. If we adopt Portuguese dredged material classification (corrected for Croatian background values) we can conclude that the worst case is found for Cu whose concentration in marine service areas classifies sediment into Class 5 — very contaminated, not to be permitted for deposition in the aquatic environment. These sediments should not be dredged, but in a case of imperious removal shall be considered as dangerous industrial residues. Following the same approach, concentrations of Zn and As in service area sediments classify these sediments to be Class 4 — contaminated and also not to be permitted for deposition in the aquatic environment. The deposition on land needs the definition of an impermeable site which will be covered by clean soil. Furthermore, sediments in marine service areas are only slightly contaminated by Pb. They belong to Class 3 – allowed for disposal into the sea but monitoring is obligating. Beaches, pelloids and delta sediments belong to Class 1 regarding all 4 elements, except Cu in category of delta sediments which classifies these sediments in Class 2 – contamination in traces. All other categories presented in Table 3 also belong to this category.

CONCLUSION

In this study the baseline data for the costal sea sediments of the eastern Adriatic Sea for elements K, Ca, Ti, Cr, Mn, Fe, Ni, Cu, Zn, Ga, As, Br, Rb, Sr, Y, and Pb have been presented. It is proposed to evaluate the natural values as a mean measured for unpolluted bays magnified by 3 standard deviations. The results of regional screening statistically evaluated and presented as distribution maps show that contamination by heavy metals Cu,

Zn, As and Pb is most pronounced. Their distribution is strongly affected by the usage of antifouling paints.

Acknowledgment

This research has been partially financed by EU FP7 Project UNCOSS, Croatian Ministry of Science, Education and Sports, Institute for Researches and Development of Defense Systems and NGO "Lijepa nasa".

References

[1] T. Cechak, M. Hlozek, L. Musılek, T. Trojek, *Nucl. Instr. and Meth.* A 580 (2007) 717.
[2] Y. Kolmogorov, V. Kovaleva, A. Gonchar, *Nucl. Instr. and Meth.* A 448 (2000) 457.
[3] M. S. Skidmore, R. M. Ambrosi, H. Simon, *Nucl. Instr. and Meth.* A 604 (2009) 592.
[4] M. West, et al., *J. Anal. At. Spectrom.* 22 (2007) 1304.
[5] M.J. Hsu, K. Selvaraj, *G.* Agoramoorthy*, Environmental Pollution* 143 (2006) 327.
[6] T. Eeva, J. Sorvari, V. Koivunen, *Environmental Pollution* 132 (2004) 533.
[7] QXAS, http://www.iaea.org/OurWork/ST/NA/NAAL/pci/ins/xrf/pciXRFdown.php [accessed on July 21, 2011]
[8] M. Juracic, C. Benac, R. Crmarić, *Geologia Croatica* 52 (1999) 131.
[9] J.H. Myers, L. Gunthorpe, G. Allison, S. Duda, *Mar. Pollut. Bull.* 52 (2006) 1048.
[10] T. Legovic, N. Limic, V. Valkovic, M. Nagj, J. *Injuk Fizika* 22 (1990) 126.
[11] S. Caeiro, M.H. Costa, T.B. Ramos, F. Fernandes, N. Silveira, A. Coimbra, G. Medeiros, M. Painho, *Ecological Indicators* 5 (2005) 151.

In: Heavy Metal Sediments
Editor: Lucía H. Sanz

ISBN: 978-1-61470-690-8
© 2012 Nova Science Publishers, Inc.

Chapter 3

ENVIRONMENTAL ASSESSMENT OF METAL CONCENTRATIONS IN SELECTED RIVERINE, ESTUARINE AND COASTAL SEDIMENTS

Yu Qian,[1] Huan Feng,[1,*] Weiguo Zhang,[2] Lizhong Yu,[2] Qingzhi Zhu,[3] Luoping Zhang,[4] and Xuchen Wang[5]

[1]Department of Earth and Environmental Studies, Montclair State University, New Jersey, USA
[2]State Key Laboratory of Estuarine and Coastal Research, East China Normal University, Shanghai, PRC
[3]School of Marine and Atmospheric Science, State University of New York, Stony Brook, NY, USA
[4]State Key Laboratory of Marine Environmental Science, Environmental Science Research Center, Xiamen University, Xiamen, Fujian, PRC
[5]Department of Environmental, Earth and Ocean Sciences, University of Massachusetts at Boston, Boston, MA, USA

ABSTRACT

Riverine, estuarine and coastal environment are important habitats for aquatic and marine life.However, long-term industrial developments with

---

*Corresponding author. Tel.: 973-655-7549; fax: 973-655-4072. E-mail: fengh@montclair.edu (H. Feng).

economic growth in coastal regions have resulted in environmental degradation, habitat loss and geomorphologic alteration in riverine, estuarine and coastal systems globally.Due to coastal development such as new industrial facilities and commercial port expansion, anthropogenic metals are introduced near the construction sites and dispersed and transported to the adjacent areas.Consequently, metal concentrations in riverine, estuarine and coastal sediments are one of the focused environmental concerns. Sediment quality reflects the long-term health status of a riverine, estuarine and coastal environment because it keeps the records of the historic development in the area.In this study, we summarize and report metal concentrations in the sediments of 43 selected sites around the world to expand our knowledge in understanding the environmental quality.The results from this study could be applicable to science-based policy formulation, and ecological restoration/rehabilitation practices in an integrated coastal management (ICM) framework to meet the urgent need for integrated coastal management planning and environmental protection.

INTRODUCTION

Metals are natural elements widely distributed in our environment in different speciation forms and oxidation state (Hill, 2010). Depending on the rock type, geophysical condition and geographical location around the world, metal concentrationspresent in sediments, soils, and water as naturally weathered products from the earth crust are different from place to place (Martin and Whitfield, 1983; Taylor and McLennan, 1995).Therefore, metal background concentrations could be very different.In general, trace amount of metals is essential micronutrients for growth of many organisms and necessary to organism metabolism (Underwood, 1971). However, metals in excessive amount can be toxic and cause pollution problem (Underwood 1971; Langston 1990). Because of human activities associated with the urbanization and industrialization in the twentieth century, anthropogenic input of metals has increased dramatically in the environment (Feng et al., 2005, 2008). For example, toxic metal pollution in estuary and coastal areas has been drawing increasing attention around the world because about 80% of the pollutants from human activities are introduced into estuary and coastal environment (Hill, 2010).Therefore, metal pollution in the estuarine and coastal sediments has become a major environmental problem because these areas are important habitats for aquatic life.

Riverine, estuarine and coastal environment are complex systems involving physical, chemical, and biological processes that play important roles in biogeochemical cycle of metals. Anthropogenic sources of metal pollution to the estuaries and coastal areas include mining, metal product fabrication, solid waste disposal, fossil fuel combustion, and municipal and industrial waste effluent (Hill, 2010). It is reported that acid mine drainage and erosion of waste dumps and tailings deposits introduce toxic metal in particulate form into aquaticsystem, whereassmelting of ore emitsmetals intoatmosphere, which end up as pollutants in soils and sediments (Salomons, 1995). Industrial wastewater is mostly from plants within the drainage area. Industries, such as foundry, paper-processing, laundry, tannery and dye works, can emit toxic metal particles and discharge wastewater intothe adjacent estuaries and coastal areas (Hunt et al., 1994; Feng et al., 1998; Onwueme and Feng, 2006). Acid rain can leach metals and increase the mobility of toxic metals from pollutant sources to the environment (Müller, 2000). Once entered aquatic environment, toxic metals canbe partitioned between dissolved phases and particulate phases with dominance in particulate phases due to their particle reactive nature.Most of these toxic metals will be ended in the sedimentary phases because of particle scavenging and settling.Therefore, high concentration of toxic metals are often found in sediments in many industrialized estuaries and coastal areas around the world because the major sources of anthropogenic metals are originated from industrialization and urban development with population growth in the areas (Forstner and Wittman, 1983; Bothner et al., 1998; Feng et al., 2004; Santos et al., 2005).In the sediment, the existence of metals can be further classified as association with exchangeable, carbonates, Fe-Mn oxides, organic matter, and residual phases as a result of the binding behavior between toxic metal and surficial sediment matrices (Filgueiras et al., 2004). For example, it is reported that depending on the metal chemical properties Zn and Cu are primarily bound to organic matter while Cr is primarily to Fe-oxides (Yu et al., 2001).It is also found that there exist significant correlations between Ni and carbonates, Cr and carbonates, Ni and Fe-oxides, Cu and Mn-oxides, and Cr and Mn-oxides, respectively (Yu et al., 2001). It is well known that fine-grained sediments are the main carriers of the toxic metals because of its higher specific surface area (Che et al., 2003). Many studies have shown that estuarine and coastal sediments are repository for metal pollutants and provide time-integrated records of pollution history (Goldberg et al. 1979; Trefry and Shokes 1981; Bopp and Simpson 1989; Valette-Silver, 1993;Feng et al., 2004, 2011).This article presentssedimentmetal concentrations in selected estuaries and coastal areas around the world, and evaluatesthe environmental impact associated with the urbanization and economic development.

Metal Concentrations in Estuarine and CoastalSediments

In this study, metal concentrations in the riverine, estuarine and coastal sedimentsarederived from the published literature covering five continents including 43 selected sites around the world. The publications are based on the field studies in many countries, such as France (Audry et al., 2004), Spain (Caliani et al., 1997; DelValls and Forja, 2002; Bermejo et al., 2003), Greece (Galanopoulou and Vgenopoulos, 2009; Pazi, 2010), Turkey (Topcuoğlu et al., 2004), Albania (Çelo et al., 1999), Italy (Buccolieri et al., 2006; Spagnoli et al., 2008), United States (Feng et al., 1998; Shumilin et al., 2002; Kim et al., 2004; Hwang et al., 2006), French Guiana (Marchand et al., 2006), Mexico (Villaescusa-Celaya et al., 2000), China (Li et al., 2000; Hong et al., 2003; Feng et al, 2011), India (Jonathan and Mohan, 2003; Reddy et al., 2004) Korea (Hyun et al., 2007), Azerbaijan (Mora et al., 2004), Iran (Mora, 2004), Kazakhstan (Mora, 2004), Russia (Mora, 2004), Fiji (Maata and Singh, 2008) and Australia (Liaghati et al., 2003) as well as Africa (Nemr et al., 2006) and Antarctica (Andrad et al., 2001).As shown in Table 1, the sediment metal concentrations vary widely from place to place within these 43 selected study areas. The recorded metal concentrations range from 0.096% - 14.4% for Al, 0.038% - 6.21% for Fe, 0.01-1763 mg kg^{-1} for Cd, 1.00-1038 mg kg^{-1} for Cr, 0.50-4280 mg kg^{-1} for Cu, 0.01-11.6 mg kg^{-1} for Hg, 0.40-32600 mg kg^{-1} for Mn, 1.20-221 mg kg^{-1} for Ni, 0.69-5778 mg kg^{-1} for Pb, and 1.0-10000 mg kg^{-1} for Zn, respectively (Table 1). The wide variations in metal concentrations reflect the different natural mineral compositions in the sediments and significant anthropogenic input in some of these coastal areas.

SedimentQualityAssessment

The environmental hazardous materials (e.g., toxic metals) can cause sediment contamination in estuaries and coastal areas.

Table 1. Metal concentrations in riverine, estuarine and coastal sediments from selected areas around the world. (Unit: mg kg $^{-1}$)

Country/ Region	Location		Al	Fe	Cd	Cr	Cu	Mn	Ni	Pb	Zn	References
Caspian Sea	Caspian Sea	Mean±S.D.	39000 ± 25000	20000 ± 15500	0.10± 0.06	56.1 ± 31.4	19.5± 16.0	482± 338	29.8 ± 21.9	11.3 ± 7.5	46 ± 38	Mora et al., 2004
		Range	1070-81300	1600-44000	0.01-0.25	1.9-128	1.2-57.6	45-1110	1.8-68.0	0.7-28.6	1-146	
Africa	Suez Gulf	Mean±S.D.		2240 ± 3789	9.92 ± 3.74	32.0 ± 18.3	33.2± 37.3	116± 204	71.4 ± 12.7	70.4± 20.3	159± 138	Nemr et al., 2006
		Range		385-13549	5.63-17.1	18.1-85.0	6.9-133	22-727	58.8-96.7	35.5-95.6	41-428	
Antarctica	Potter Cove	Mean±S.D.		132100 ± 23254		6.3± 0.6	117± 23	957± 86		3.7 ± 0.3	52 ±1	Andrad et al., 2001
		Range		108600-155100		5.7-6.7	94-141	880-1050		3.4-3.9	51-54	
Australia	Bells Creek catchment	Range	955-9821	9400-41000		1.0-26.0	0.5-5.8	0-229	1.5-6.4	3.0-7.3	18-68	Liaghati et al., 2003
China	Mingjiang Estuary	Mean±S.D.			0.16± 0.08		27.9 ± 7.4			43.8 ± 11.9	100 ± 13	Hong et al., 2003
		Range			0.09-0.34		16.4-37.2			34.4-69.6	92 – 128	
	Pearl River estuary	Mean±S.D.					37.7 ± 1.7		32.3 ± 0.9	50.9 ± 3.3	104 ± 7	Li et al., 2001
		Range					33.9-40.4		31.2-34.6	47.0-59.5	95-115	
	Port of Tianjin	Mean±S.D.		34366 ± 4948	0.14 ± 0.04	55.9 ± 6.4	31.0± 10.1		31.8 ± 2.5	21.9 ± 2.4	91 ± 19	Feng et al, 2011

Table 1. (Continued)

Country/ Region	Location		Al	Fe	Cd	Cr	Cu	Mn	Ni	Pb	Zn	References
		Range		24615-38900	0.11-0.24	48.0-73.0	25.9-62.4		28.1-35.3	19.2-26.1	75-147	
	Xiamen-Jinmen	Mean±S.D.			0.21±0.05		16.6±4.2			52.6±14.5	104±26	Hong et al., 2003
		Range			0.14-0.29		11.0-24.5			36.0-80.3	78-161	
	Yangtze River intertidal	Mean±S.D.	65375±11467	33394±6257	0.26±0.13	78.9±19.7	30.7±9.7	766±177	31.8±7.2	27.3±5.6	94±24	Zhanget al., 2009
		Range	40803-97213	20538 49627	0.12-0.75	36.9-173	6.9-49.7	413-1112	17.6-48.0	18.3-44.1	48-154	
Europe	Adriatic Albanian coast	Mean±S.D.		44060±14809	0.22±0.13	313±52	103±92	1940±2364	240±65	18.4±6.4	107±58	Babi et al., 1999
		Range		24600-61100	0.14-0.45	263-401	31.6-255	499-6149		13.1-29.2	48-193	
Fiji	Suva harbour	Mean±S.D.		31533±9300			54.0±29.8			45.4±17.4	110±48	Maata et al., 2008
		Range		14000-48700			21.4-143			22.1-93.5	40-269	
France, Lot River	Cajarcsite,Lot River	Mean			125		97.7			523	4430	Audry et al., 2004
		Range			12.6-294		40.6-264			111-1280	909-10000	
	Marcenacsite, Lot River	Mean			0.81		269			105	134	
		Range			0.33-1.40		17.9-42.0			27.9-68.0	82-206	

Country/ Region	Location		Al	Fe	Cd	Cr	Cu	Mn	Ni	Pb	Zn	References
	Templesite,Lot River	Mean			20.4		30.7			105	949	
		Range			0.56-180		16.7-78.0			31.3-593	82-7230	
French Guiana	Kaw	Mean±S.D.		47245±4077		55.6±8.3		529±160	35.2±4.7	26.9±4.1	174±28	Marchand et al., 2006
		Range		39762-59922		39.5-67.1		240-901	27.0-48.7	16.6-55.9	128-267	
	Shoreface sediments	Mean±S.D.		50144±558		64.5±1.0	24.8±1.9	1040±40	35.8±1.8	30.0±2.1	141±15	
		Range		49530-51317		63.4-65.5	22.9-26.7	1001-1089	34.0-37.6	26.9-33.2	123-154	
	Sinnamary	Mean±S.D.		44062±5361		59.8±11.4	17.8±6.4	539±509	31.7±5.9	26.9±4.1	164±71	
		Range		24628-62993		31.7-72.8	3.8-38.8	1198-2494	18.8-44.6	16.6-37.3	82-388	
Greece	Candarli Gulf	Mean±S.D.	18241±6438	26781±5259		41.2±12.7	24.4±10.7	400±256	42.0±20.8	35.5±20.7	94±53	Pazi et al., 2010
		Range	3800-26800	16200-33800		15.8-66.7	2.7-46.1	167-1423	7.6-92.7	14.5-138	50-358	
	Keratsini Harbor, Saronikos Gulf	Mean±S.D.			998±329	464±145	288±75	204±185		648±151	1435±1438	Galanopoulou et al., 2009
		Range			190-1763	264-860	195-518	95-1101		521-1263	409-6725	
India	Alang–Sosiya coast	Mean±S.D.	92100±24927	76304±74672	30.2±19.4	203±96	179±119	6167±5985	124±85	156±99	1140±389	Reddy et al., 2004
		Range	73400-120400	26749-162189	8.6-45.9	137-313	85,2-313	1751-12979	60.3-222	66.0-138	718-1483	
	Gulf of Mannar	Mean±S.D.		12422±7246	0.20±0.06	183±199		301±169	23.7±11.4	15.7±3.5	74±27	Jonathan et al., 2003

Table 1. (Continued)

Country/ Region	Location		Al	Fe	Cd	Cr	Cu	Mn	Ni	Pb	Zn	References
Italy	Gulf of Manfredonia	Range		3600-34600	0.06-0.35	34-1038		70-818	11.0-63.0	7.0-21.0	41-184	
		Mean±S.D.				81±24	47±30		47±15	17±6	82±21	Spagnoli et al., 2008
		Range				15-122	11-135		12-82	4-33	23-119	
	Naples city port	Mean±S.D.			0.90±0.66	72.5±42.7	131±84	389±94		123±68	303±274	Adamo et al., 2005
		Range			0.20-2.50	10.3-162	40-415	95-535		37-314	41-1196	
Southern Italy	Taranto Gulf	Mean±S.D.	58564±11247	31566±2903		85.9±7.7	47.4±2.7	893±537	53.3±3.8	57.8±8.4	102±14	Buccolieri et al., 2006
		Range	36866-72020	26313-36098		75.2-103	42.4-52.3	552-2826	47.9-60.7	44.7-74.8	87-129	
Korea	Masan Bay	Mean±S.D.		31566	1.24	67.1	43.4		28.8	44	206	Hyun et al., 2007
		Range		26313-36098	0.10-7.47	30.5-99.8	13.5-90.7		10.2-40.4	13.0-82.2	80-379	
Mexico, U.S.	Baja California, California	Mean±S.D.	57000±14927	27200±7200	0.17±0.12	137±165	12,8±3,9	889±1096	25.3±8.0	13.6±3.8	68±32	Villaescusa et al., 2000
		Range	36100-85500	13600-46000	0.08-0.64	55.9-802	4.9-22.7	392-5282	15.5-44.4	6.0-21.3	39-188	
Moroccan	Lagoon of Oualidia	Mean±S.D.	108600±25400	69100±8900		52.5±6.1	36.5±22.9			54.6±30.6	228±16	Zourarah et al., 2007
		Range	52600-143600	46200-79000		46.0-62.0	20-90			88.0-16.6	203-260	

Country/ Region	Location		Al	Fe	Cd	Cr	Cu	Mn	Ni	Pb	Zn	References
Spain	Estuary of Huelva and adjacent Atlantic shelf	Mean±S.D.			3.25± 3.06	52.0± 39.9	407± 603		21.2± 14.9	191± 283	535± 694	Caliani et al., 1997
		Range			1.- 9	16- 150	17- 1830		10- 46	12- 926	61- 2300	
	Gulf of Cadiz	Mean±S.D.		27910± 11911	0.80± 0.24	87.3± 65.4	53.4± 19.8	293± 87	29.4± 12.3	45.0± 21.5	118± 52	DelValls et al., 2002
		Range		1271- 39820	0.39- 1.24	40- 284	13.2- 92.4	79- 452	1.2- 43.8	6.9- 84.6	19- 188	
	Odiel River	Mean±S.D.		31862± 21718		54.7± 52.4	604± 751	1086± 1212	29.7± 17.4	2369± 1925	2874± 1647	Bermejo et al., 2003
		Range		11601- 88429		7.9- 34.6	52- 2843	54- 5482	15.2- 95.1	512- 5778	1264- 8286	
Turkey	Marmara Sea	Mean±S.D.		10909± 4548	0.05± 0.06	47.8± 18.1	20.0± 9.4	344± 61	38..6 ± 16.8	25.4± 5.7	43 ±8	Topcuoğlu et al., 2004
		Range		5956- 14896	0.02- 0.12	27.2- 61.5	12.7- 30.6	274- 384	20.5- 53.9	21.6- 31.9	34-51	
U.S	Hudson River	Mean±S.D.	51264± 5980	34348± 5598	1.00± 0.56		77.8± 31.7			81.5± 33.5	182± 41	Feng et al., 1998
		Range	41000- 63200	20300- 42800	0.18- 2.29		18.0- 149			24.0- 177	101-	
	California, Tidal salt marsh	Mean±S.D.	27220± 8218	32680± 10314	0.46± 0.11	70.3± 19.6	37.9± 26.6		53.4± 21.9	39.1± 33.0	120± 87	Hwang et al., 2006
		Range	13700- 35500	17200- 42600	0.18- 0.26	18.3- 59.1	15.1- 78.3		20.3- 75.1	6.8- 87.9	67- 273	
	Upper Gulf of California	Mean±S.D.		17298± 8091		28.1± 13.0			45.1± 44.0		44± 27	Shumilin et al., 2002
	Colorado River Delta	Mean±S.D.		16740± 6138		27.6± 8.8			117± 117		39 ±9	Shumilin et al., 2002

Table 1. (Continued)

Country/ Region	Location		Al	Fe	Cd	Cr	Cu	Mn	Ni	Pb	Zn	References
	Oyster Rock Landing salt marsh in Delaware	Mean±S.D.	46067± 6754				29.6± 4.7			36.1± 11.8	66±18	Kim et al., 2004
		Range	43000- 55900				23.0- 38.0			19.0- 55.0	42-89	
	Upper Gulf of California	Mean±S.D.		17600± 102400	0.30± 1.90		25± 4820	50± 32600	13.0± 190	11± 380	100± 3540	Kim et al., 2004
	Wolfe Glade Delaware salt marsh	Mean±S.D.	28878± 7017				37.8± 36.8			130± 74	78± 14	Kim et al., 2004
		Range	17800- 38700				13- 130			22- 215	51- 100	
Earth Crust			80400	35000	0.098	35	25	600	20	20	71	Taylor and McLennan, 1995

Therefore, metal enrichment factor has been widely used as the assessment criteria to study the sources and contamination of toxic metals in estuarine and coastal environment to prevent and control environmental pollution, protect marine life, ensure natural resource sustainable uses, maintain ecological equilibrium, and protect human health (e.g., Salomons and Forstner, 1984; Sinex and Wright, 1988; Windom et al., 1989, 1991; Feng et al., 1998, 2004, 2011; Zhang et al., 2009; Zhang and Liu, 2002). In this section, estuarine and coastal sediment quality is evaluated by metal enrichment factor (EF).Metal enrichment factors (EF) are commonly used to identify metal concentrations of environmental concern (GIPME, 1999). It is widely used in sediment quality assessment (e.g., Feng et al., 1998, 2004, 2011;GIPME, 1999). Based on the EF values, natural metal concentrations can be distinguished from those of anthropogenic origin (Sinex and Wright, 1988; Windom et al., 1989; Schropp et al., 1990; Zhang and Liu, 2002). Therefore, EF is a very useful indicator to evaluate sediment quality and metal contamination.Metal EF is mathematically defined by the following equation (e.g., Sinex and Helz, 1981):

$$EF = \frac{\left(\frac{Me}{Re}\right)_{Sample}}{\left(\frac{Me}{Re}\right)_{Background}} \qquad (1)$$

where Me is the metal concentration of concern, Re is the concentration of either Al or Fe that are the most abundant element in the earth crust (Martin and Whitfield, 1983; Taylor and McLennan,1995) , $(Me/Re)_{Sample}$ is the ratio in the sample, and $(Me/Re)_{Background}$ is the ratio of the background. Ideally, the metal background concentration should be derived from the sampling site (Kersten and Smedes, 2002; Liu et al., 2003; Covelli et al., 2006). In most of the cases, however, the background data are not readily available.In this case, metal concentrations in the upper continental crust (Taylor and McLennan, 1995) are alternatives for thebackground values.The same approach has been used in our previous studies (e.g., Feng et al., 2004; Zhang et al., 2007, 2009). For metal enrichment factor calculation performed in this study, we use the metal background values wherever they areavailable in the local study area.When the local metal background values arenot available, we adopt the upper continental crust value as the background. Also, we haveto use either Al or Fe in the EF calculation because Al data arenot always available.However, our method is still valid because Al and Fe are commonly used for EF calculation (Quevauviller et al., 1989; Isakson et al., 1997; Lee et al., 1998; Reimann and de Caritat, 2000; Bergamaschi et al., 2002; Hernandez et al., 2003; Conrad and Chisholm-Brause, 2004; Mishra et al., 2004).The available Al and Fe data from this study demonstrate a significant correlation (r=0.7745, n=13) between Al and Fe in the sediments (Figure 1), which supports our adoption.

Metal EFs provide an indication of the metal sources of and enrichment stage in the sediments.The assessment criteria using EF are generally based on the EF values (Salomons and Forstner, 1984; Sinex and Wright, 1988; Klerks and Levinton, 1989; Sutherland et al., 2000; Zhang and Liu, 2002; Loska and Wiechu1a, 2003; Yongming et al., 2006).In a simple classification, EF ≈ 1 indicates natural crustal origin, whereas EF>10 suggests anthropogenic source (Nolting et al., 1999).Zhang and Liu (2002) also recommended that 0.5 < EF < 1.5 suggest that the metals may be entirely from natural crustal weathering processes, whereas EF > 1.5 suggests that a significant portion of the metal is delivered from non-crustal materials. In a more detail classification, metal EF values can be divided into five categories based on the degree of enrichment, i.e., 1) EF < 2 suggests deficiency to minimal enrichment; 2) EF = 2–5, moderate enrichmen; 3) EF = 5–20, significant enrichment; 4) EF = 20–40, very high enrichment; and 5) EF > 40, extremely high enrichment (Yongming et al., 2006).

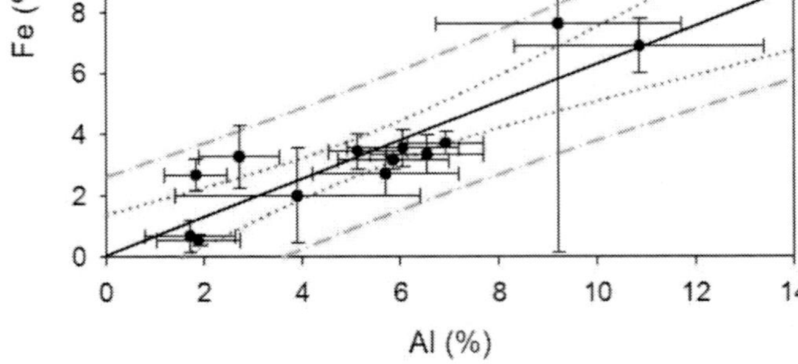

Figure 1.A significant correlation (p< 0.001) between Al and Fe in the sediments. Data are selected from 13 sites where both Al and Fe concentrations are available.

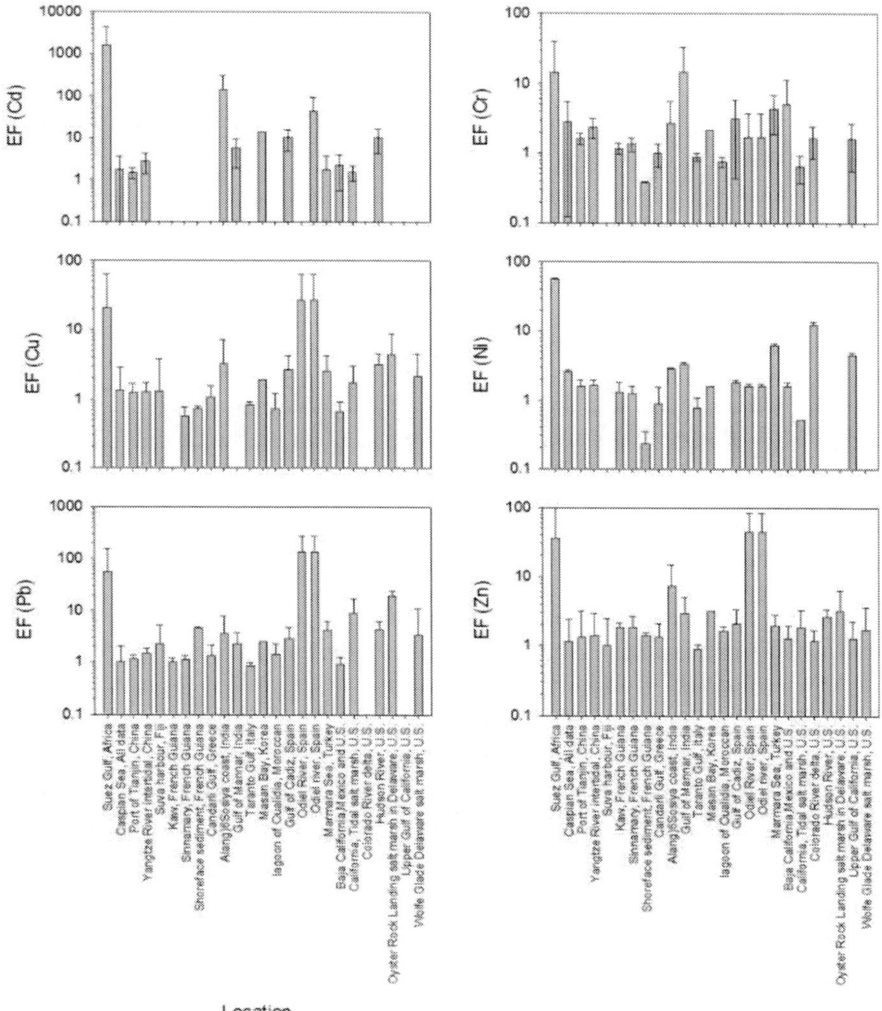

Figure 2. Enrichment factors (EF) of Cd, Cr, Cu, Ni, Pb and Zn in the sediments. The wide variations in metal enrichment factors among these sites reflect the different sources and enrichment of metals. High enrichment (EF > 2) suggests various degrees of metal pollution.

Our calculation of metal enrichment factor (EF) has to depend on the data availability in the area.Therefore, there may exist some unintentional bias in our dataset during the screen and selection.Nevertheless, the results from 24 selected estuaries and coastal areas for this study show that metal EF values vary widely from location to location from deficiency to extremely high enrichment (Figure 2). Specifically, the enrichment factors range from 0.2 – 2741 for Cd, 0.10 – 84 for Cr; 0.08 – 125 for Cu, 0.10 – 19 for Mn, 0.08 – 76 for Ni, 0.06 – 317 for Pb and 0.2 – 128 for Zn, respectively (Figure 2).Further analysis shows that among the metals in the study areas, some of them show different degree of enrichment as summarized below:

1) Cd – 9 out of 13 sites show unnatural enrichment, of which Suez Gulf, Alang-Sosiya coast and Odiel River show extremely high enrichment.
2) Cr –9 out of 20 sites show enrichment accumulation, of which Suez Gulf and Gulf of Mannar show significant enrichment level.
3) Cu – 8 out of 20 sites show sign of toxic metal enrichment, of which Suez Gulf and Odiel River show very high enrichment.
4) Ni – 7 out of 19 sitesshow metal enrichment, of which only Suez Gulf shows extremely high enrichment.
5) Pb – 13 out of 22 sitesshow unnatural enrichment, of which Suez Gulf and Odiel River show extremely high enrichment while Oyster Rock Landing salt marsh in Delaware shows significant enrichment.
6) Zn – 8 out of 24 sites show Zn enrichment, of which Odiel River shows extremely high enrichment while Suez Gulf shows very high enrichment.

As shown in Table 2, cases of moderate to high enrichment (EF > 2) accounts for 33% - 69%. Relatively high percentages of enrichment occur to Cd (69%) and Pb (59%).Our study shows that metal enrichment factors reflect the sediment quality and the influence of local point sources of contaminants. Although we are not able to identify specific sources of these contaminants, we are certain that these contaminations are caused by local anthropogenic activities related to economic development.

Table 2. Statistical analysis of metal enrichment factor in the selected estuaries and coastal areas around the world, and the percentages of the cases with EF>2

Metal	Cd	Cr	Cu	Ni	Pb	Zn
n of cases studied	13	20	20	19	22	24
Minimum EF value	1.5	0.39	0.57	0.23	0.82	0.88
Maximum EF value	1582	15	27	56	130	44
% of case with EF>2	69	45	40	37	59	33

Metal Ecotoxicological Assessment

Several approaches have been often used to assess the metal hazard potential to organisms, causing adverse biological effects (Urban and Cook, 1986; Long et al., 1995, 1998). The potential ecotoxicological risk on organisms in the sediments can be evaluated using sediment guidelines (e.g., Long and Morgan, 1990; Long et al., 1995, 1998; Ingersoll er al., 1996; Sundaray et al., 2011). Long et al. (1995) defined the effect range-low (ERL) and effect range-median (ERM) system based on the compilation of matching biological and chemical data from numerous modeling, laboratory, and field measurement in marine and estuarine sediments in order to better predict the toxicity of contaminants. According to Long et al. (1995), ERL is defined based on the calculation of the 10th percentiles of organism influenced by the toxicity of a specific contaminant, while ERM value is based on the 50th percentiles of organism influenced by the toxicity of that contaminant. Three effect ranges of contaminant level canbe set up by ERL and ERM. When the contaminant concentration is below ERL, a rarely biological adverse effect is anticipated, while the concentration is between ERL and ERM, occasional biological adverse effect is anticipated. However, when the concentration is greater than ERM value, a frequent biological adverse effect (> 50% chances) could occur (Long et al., 1995, 1998; Ingersoll et al., 1996). To evaluate the potential biological adverse effect, hazard quotient (HQ) is one kind of single-value estimate, and is also the simplest method to estimate toxicity potential of the selected contaminant in the sediments (Urban and Cook, 1986; Wang et al., 2002). According to Urban and Cook (1986), the HQ is estimated by the following equation:

$$\mathrm{HQ} = \frac{\mathrm{SCC}}{\mathrm{SQG}} \tag{2}$$

where SCC is the sediment chemical concentration while SQG stands for the concentration defined by the sediment quality guideline. Both SCC and SQG are in mg kg^{-1}. Since ERL is more reasonably predictive of non-toxic condition, the SQG is set at ERL levels: Cd=1.2, Cr=81, Cu=34, Ni=20.9, Pb=46.7 and Zn=150 (Long et al., 1995). In fact, HQ indicates the potential risk to organisms.If HQ<0.1, no adverse effects are expected; at $0.1 < HQ < 1$, low potential hazards are expected; in the range of $1.0<HQ<10$, some adverse effects or moderate hazards are probable; and when HQ>10, high hazard potential is anticipated (Wang et al., 2002; Finger et al., 2004)

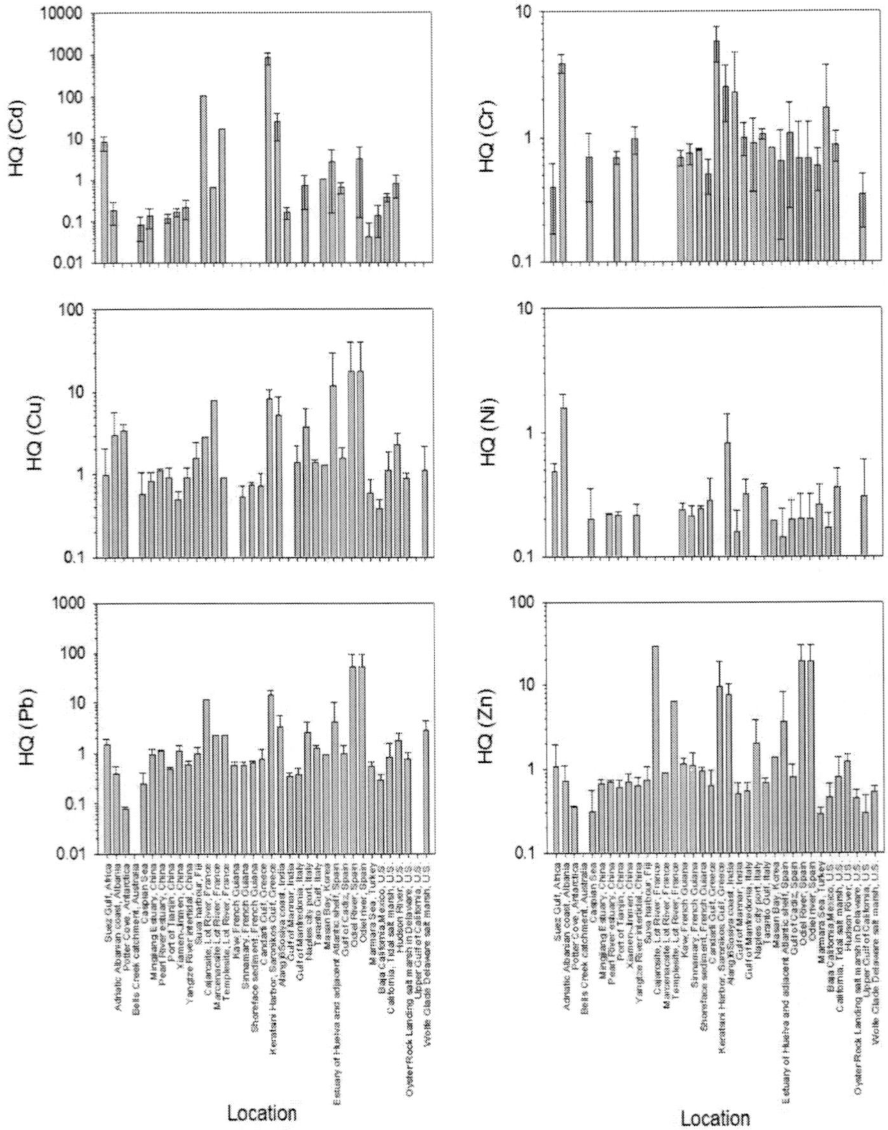

Figure 3.Hazard quotients (HQ) of Cd, Cr, Cu, Ni, Pb and Zn in the sediments. Moderate (1 < HQ < 10) to high (HQ > 10) hazard potential to cause adverse biological effect are anticipated in some riverine, estuarine and coastal areas.

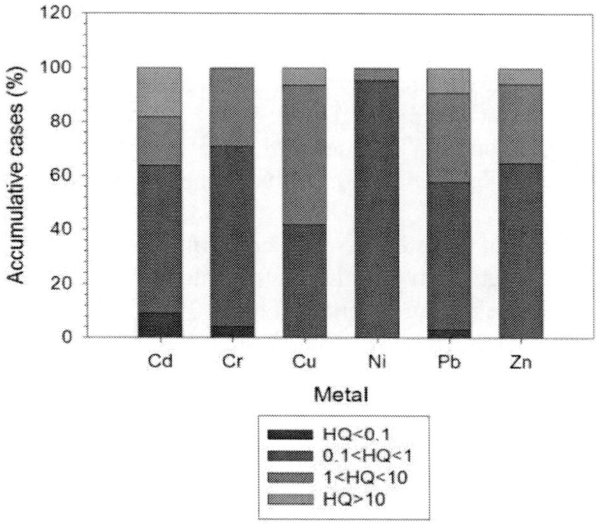

Figure 4. Percentage of riverine, estuarine and coastal areas that have no (HQ < 0.1), low (0.1 < HQ < 1), moderate (1.0<HQ<10) and high (HQ > 10) potential hazards based on calculated HQ values of Cd, Cr, Cu, Ni, Pb and Zn. Cases vary from metal to metal due to the availability of the metal data. Total cases of each individual metals are $n_{Cd} = 22$, $n_{Cr} = 24$, $n_{Cu} = 31$, $n_{Ni} = 22$, $n_{Pb} = 33$, and $n_{Zn} = 34$, respectively.

As shown in Figure 3, we found in this study that HQ values range from 0.04 – 832 for Cd (n=22), 0.08 – 5.7 for Cr (n=24), 0.38 – 17.8 for Cu (n=31), 0.14 – 1.6 for Ni (n=22), 0.08 – 50.7 for Pb (n=33), and 0.29 – 29.5 for Zn (n=34). The percentages of the estuaries and coastal areas facing various degree of hazard potential from no adverse effects (HQ < 0.1) to high hazard potential (HQ > 10) are shown in Figure 4.In most of the areas, only low hazard potentials (0.1< HQ < 1) caused by these selected metals are expected.Except for Cd, high hazard potentials (HQ > 10) caused by Cr, Cu, Ni, Pb and Zn are expected in <10% of the study areas (Figure 4).Specifically, cases of potential adverse biological effect caused by each individual metal are described below:

1) Cd –8 out of 22 sites have Cd concentrations that exceed the ERL value and suggest moderate to high hazard potential, of whichSuez Gulf, Cajarcsite, Lot River, Keratsini Harbor and Alang-Sosiya coastshow high hazard potential caused by Cd (Figure 3).
2) Cr – 5 out of 24 sites (i.e., Adriatic Albanian coast, Keratisini Harbor, Alang-Sosiya coast,Gulf of Mannar, and Baja California) have Cr concentration exceeding the ERL value and show moderate hazard

potential (Figure 3).

3) Cu – In addition to 2 out of 31 sites (i.e., Estuary of Huelva and Odiel River) have high hazard potential, another 16 out of 31 sites have moderate hazard potential (Figure 3).
4) Ni –only 1 out of 22 sites (i.e., Adriatic Albania coast) has Ni concentration exceeding ERL value and has moderate hazard potential (Figure 3).
5) Pb – 14 out of 33 sites have Pb concentration exceeding ERL value, of which 3 sites have high hazard potential while another 11 sites have moderate hazard potential (Figure 3).
6) Zn –12 out of 34 sites have Zn concentration exceeding ERL value, of which 2 sites have high hazard potential while another 10 sites have moderate hazard potential (Figure 3).

Discussion and Conclusion

Toxic metal concentrations in the sediments from the most areas selected forthis study are within the range of the natural levels. However, industrial activities, harbors, mining, wastewater and other anthropogenic sources are the main sources of toxic metal pollutants. As shown in this study, the concentrations of toxic metals in sediment are highly related to the pointsourcesinsurrounding environment.For example, Suez Gulf received inputs of toxic metals mostly from municipal, dock and industrial activities (Nemr et al., 2006),causing very high toxic metal enrichment in Suze Gulf (Figure 2).However,only Cd had exceeded ERL value (Figure 3).This is because the sandy texture bottom sedimenthas limited capacity of absorbing toxic metals by(Nemr et al., 2006). Odiel River received toxic metal inputs from industrial and mining activities (Bermejo et al., 2003), causingvery high toxic metal enrichment and biological toxicity of Cu, Pb and Zn. Thiscould be attributed to the sediment acidity (i.e., pH insediment)controllingthe toxic metalsin this areabecausethe accumulation of toxic metals increases significantly with the increasing pH (Bermejo et al., 2003). It was also found that Pb and Zn concentrations increased with the increasing carbonate concentration in the estuary although the dominant texture of sediments was sandy materials (Bermejo et al., 2003).Keratsini Harbor and Saronikos Gulf received toxic metal from sewage outfall, industrial and ship contaminants (Galanopoulou et al., 2009).Toxic metals (Cu, Cd, Zn, Pb and Cr) were highly enriched in these sediments and exceeded the ERL value (Figures 2 and 3), suggesting this area could be highly toxicto organism (Galanopoulou et al.,

2009). Lot River water system receivedtoxic metal pollution from mining and smelting activities50 years ago (Audry et al., 2004). Even though the pollution source was no longer there later, the residual Cd, Pb and Zn concentrations in the sediments were still higher than the ERL values and this river system was classified as a severely polluted area because toxic metals were still being leached out from the former smelting sites (Audry et al., 2004). As one of the largest ship scrapping work in the world, Alang-Sosiya stretches about 14 km^2 and received pollutants including toxic metals, petroleum hydrocarbons, domestic wastes and bacterial contaminants (Reddy et al., 2004). Bothenrichment factors (EFs) and hazard quotients (HQs) of Cd and Zn in this area were above natural levels (Figures 2 and 3). High Cd and Zn concentrations in the sediments could be mainly attributed to the ship scrapping industry in this area (Reddy et al., 2004). The Gulf of Mannar is located in an area with high marine fauna production, and received pollutants from industrial sites, Tuticorin Harbor and River Tambraparani during the past three decades (Jonathan et al., 2003). The EF and HQ of Cr show significant enrichment and moderate hazard potential in this area (Figures 2 and 3).High Cr concentration in the sediments could be attributed to heavy industrial activities in this area as well as carbonate and skeletal fragment being the main carrier phase (Jonathan et al., 2004). In Adriatic Albanian coast, pollutants were transported from industrial activities such as mining, commercial harbor and chlor-alkali plant, causing high Cr and Ni concentrations (Babi et al., 1999). Both Cr and Ni hadmoderate hazard potential to the organisms in this area (Figure 3).

Based on the analysis of the information collected for this study, our results suggest that toxic metal contaminations willcontinue to raise our concerns because metal contaminations are still present in the world's estuaries and coastal areas.The current tasks facing to usare how we can effectively exercise the environmental protection, ecosystem restoration, and sustainable development of the coastal areas. These are also the worldwide issues becausethey are not just restricted in any localized area although the specific focuses could be different to a certain extent.

REFERENCES

Adamo, P., Arienzo, M., Imperato, M., Naimo, D., Nardi, G., Stanzione, D. (2005) Distribution and partition of heavy metals in surface and sub-surface sediments of Naples city port. *Chemosphere*, 61, 800-809.

Andrade, S., Poblet, A., Scagliola, M., Vodopivez, C., Curtosi, A., Pucci, A., Marcovecchio, J. (2001) Distribution of heavy metals in surface sediments from an Antarctic marine ecosystem.*Environmental Monitoring and Assessment,* 66, 147-158.

Audry, S., Schäfer, J., Blanc, G., Jouanneau, J. M. (2004) Fifty-year sedimentary record of heavy metal pollution (Cd, Zn, Cu, Pb) in the Lot River reservoirs (France). *Environmental Pollution*, 132, 413-426.

Bergamaschi, L, Rizzio, E, Valcuvia, M. G., Verza, G., Profumo, A., Gallorini, M. (2002) Determination of trace elements and evaluation of their enrichment factors in Himalayan lichens.*Environmental Pollution*, 120, 137–44.

Bermejo, S. J. C., Beltrán, J. L., Ariza, G. (2003) Spatial variations of heavy metals contamination in sediments from Odiel river (Southwest Spain). *Environment International*, 29, 69-77.

Bopp, R. F. and Simpson, H. J. (1989) Contamination of the Hudson River - The sediment record. In: Contaminated Marine Sediments Assessment and Remediation. pp. 401-416.

Bothner, M. H., Buchhltz ten Brink, M. Manheim, F. T. (1998) Metal concentrations in surface sediments of Boston Harbor – Changes with time. *Marine Environmental Research*, 45, 127-155.

Buccolieri, A., Buccolieri, G., Cardellicchio, N., Dell'Atti, A., Leo, A. D., Maci, A. (2006) Heavy metals in marine sediments of Taranto Gulf (Ionian Sea, Southern Italy). *Marine Chemistry*, 99, 227-235.

Caliani, F. J. C., Muñoz, F. R. Galán, E. (1997) Clay mineral and heavy metal distributions in the lower estuary of Huelva and adjacent Atlantic shelf, SW Spain. *The Science of the Total Environment*, 198, 181-200.

Çelo, V., Babi, D., Baraj, B., Çullaj, A. (1999) An assessment of heavy metal pollution in the sediments along the Albanian coast. *Water, Air, and Soil Pollution*, 111, 235-250.

Che, Y., He, Q., Lin, W. Q. (2003) The distributions of particulate heavy metals and its indication to the transfer of sediments in the Changjiang Estuary and Hangzhou Bay, China. *Marine Pollution Bulletin*, 46, 123-131.

Conrad, C. F., Chisholm-Brause, C. J. (2004) Partial survey of trace metal contaminations in the sediments of the Elizabeth River, Virginia. *Marine Pollution Bulletin*, 49, 319–24.

Covelli, S., Fontolan, G., Faganeli, J., Ogrinc, N. (2006) Anthropogenic markers in the Holocene stratigraphic sequence of the Gulf of Trieste (northern Adriatic Sea).*Marine Geology*, 230, 29-51.

DelValls, T. A., Forja, J. M., Gómez-Parra, A. (2002) Seasonality of contamination, toxicity, and quality values in sediments from littoral ecosystems in the Gulf of Cádiz (SW Spain). *Chemosphere*, 46, 1003-1043.

Feng H., Cochran J. K., Lwiza H., Brownawell B., Hirschberg D. J. (1998)Distribution of Heavy Metal and PCB contaminants in the sediments of an urban estuary: the Hudson River.*Marine Environmental Research*, 45, 69-88.

Feng H., Jiang H., Gao W., Weinstein M. P., Zhang Q., Zhang W., Yu L., Yuan D., Tao J. (2011)Metal contamination assessment for sediments in western Bohai Bay and adjacent urban estuaries, China. *Journal of Environmental Management*, 92, 1185-1197.

Feng H., Yu L., and Solecki W. (2005). URBAN DIMENSIONS OF ENVIRONMENTAL CHANGE——*Science, Exposure, Policies, And Technologies*.Science Press USA Inc., Monmouth Junction, New Jersey, USA.324 pp.

Feng, H., Han, X., Zhang, W., Yu, L. (2004) A preliminary study of heavy metal contamination in Yangtze River intertidal zone due to urbanization. *Mar. Pollut. Bull.* 49, 910-915.

Feng, H., Zhang, W., Zhang, L., Wang, X.-C., Yu, L. and Yu, D. (2008) Chapter 9. Heavy metal contamination in selected urban coastal regions in US and China.In: Causes and Effects of Heavy Metal Pollution.(edsMikel L. Sanchez). Hauppauge, NY: Nova Science Publishers, Inc.pp. 265-286.

Filgueiras, A. V., Lavilla, I., Bendicho, C. (2004) Evaluation of distribution, mobility and binding behavior of heavy metals in surficial sediments of Louro River (Galicia, Spain) using chemometric analysis: a case study. *Science of the Total Environment*, 330, 115-129

Forstner, U., Wittman, G. T. W. (1983). Metal pollution in the aquatic environment, 2nd.Ed., Verlag Publishers. New York. 18-19.

Galanopoulou, S., Vgenopoulos, A., Conispoliatis, N. (2009) Anthropogenic Heavy Metal Pollution in the Surficial Sediments of the Keratsini Harbor, Saronikos Gulf, Greece. *Water Air Soil Pollution*, 202, 121-130.

GIPME (Global Investigation of Pollution in the Marine Environment). (1999) Guidance on Assessment of Sediment Quality.IOCeUNEPeIMO, 23 pp.

Goldberg, E. D., Griffin, J. J., Hodge, V., Koid, M., Windom, H. L. (1979) Pollution history of the Savannah River estuary.*Environmental Science and Technology*, 13, 588-594.

Hernandez, L, Probst, A., Probst, J. L., Ulrich, E. (2003) Heavy metal distribution in some French forest soils: evidence for atmospheric contamination. *Science Total Environment*, 312, 195–219.

Hill, M. K. Understanding Envrionmental Pollution. New York: Cambridge, 2010, eBook.

Hong, L.Y., Hong, H. S., Chen, W. Q., Wang, X. H., Zhang, L. P. (2003) Heavy metals in surface sediments from Minjiang Estuary-Mazu and Xiamen-Jinmen sea area. *Journal of Environmental Sciences*, 15, 116-122.

Hunt, C. D., West, D. A., Lewis, D. A. (1994) Trace metals concentrations in New York/New Jersey Harbor. In: Hazardous and Industrial Wastes. (ed. C. P. Huang). pp. 691-698.

Hwang, H. M., Green, P. G., Higashi, R. M., Young T. M. (2006) Tidal salt marsh sediment in California, USA. Part 2: Occurrence and anthropogenic input of trace metals. *Chemosphere*, 64, 1899-1909.

Hyun, S., Lee, C. H., Lee, T., Choi, J. W. (2007) Anthropogenic contributions to heavy metal distributions in the surface sediments of Masan bay, Korea.*Marine Pollution Bulletin*, 54, 1059-1068.

Ingersoll, C. G., Haverland, P. S., Brunson, E. L. (1996) Calculation and evaluation of sediment effect concentrations for the amphipod hyalellaazteca and the midge Chironomusriparius. *Journal of Great Lakes Research,* 22, 602-623.

Isakson, J., Oblad, M., Lindgren, E. S., Fridell, M. D., Pacyna, J. M., Mäkinen, M. (1997) Perturbation of background aerosol at rural sites in the Nordic countries.*Atmosphere Environment*, 31, 3077–3086.

Jonathan, M. P., Mohan, V. R. (2003) Heavy metals in sediments of the inner shelf off the Gulf of Mannar, South East Coast of India. *Marine Pollution Bulletin*, 46, 258-268.

Kersten, M., Smedes, F. (2002) Normalization procedures for sediment contaminants in spatial and temporal trend monitoring.*Journal of Environment Monitoring*, 4, 109-115.

Kim, G., Alleman, L.Y., Church, T. M. (2004) Accumulation records of radionuclides and trace metals in two contrasting Delaware salt marshes. *Marine Chemistry*, 87, 87-96.

Klerks, P. L., Levinton, J. S. (1989). Rapid evolution of metal resistance in a benthicoligochaete inhabiting a metal-polluted site.*The Biological Bulletin*, 176, 135-141.

Langston, W. J. (1990). Toxic effects of metal and the incidence of metal pollution in marine ecosystem. In: Heavy metals in the marine Environment (Eds. Furness R.W. and Rainbow P.S.). CRC Press, Boca Raton, 101-1022.

Lee, C. L., Fang, M. D., Hsieh, M. T. (1998) Characterization and distributing of metals in surficial sediments in southwestern Taiwan. *Marine Pollution Bulletin*, 36, 464 – 471.

Li, X., Wai, O. W. H., Li, Y. S., Coles, B. J., Ramsey, M. H., Thornton, I. (2000) Heavy metal distribution in sediment profiles of the Pearl River estuary, South China. *Applied Geochemistry*, 15, 567-581.

Liaghati, T., Preda, M., Cox, M. (2003) Heavy metal distribution and controlling factors within coastal plain sediments, Bells Creek catchment, southeast Queensland, Australia. *Environment International*, 29, 935-948.

Liu, W. X., Lia, X. D., Shen, Z. G., Wang, D. C., Wai, O. W. H., Li, Y. S. (2003) Multivariate statistical study of heavy metal enrichment in sediments of the Pearl River Estuary. *Environmental Pollution*, 121, 377-388.

Long, E. R., Field, L. J., MacDonald, D. D. (1998) Predicting toxicity in marine sediments with numerical sediment quality guidelines. *Environmental Toxicology Chemistry*, 17, 714-727.

Long, E. R., Macdonald, D. D., Smith, S. L., Calder, F. D. (1995) Incidence of adverse biological effects within ranges of chemical concentrations in marine and estuarine sediments. *Environmental Management*, 19, 81-97.

Long, E. R., Morgan, L. G., (1990) The Potential for Biological Effects of Sediment-sorbed Contaminants Tested in the National Status and Trends Program.NOAA Technical Memorandum NOS OMA 52.National Oceanic and Atmospheric Administration, Seattle, Washington, 175 pp.

Loska, K., Wiechula, D. (2003) Application of principal component analysis for the estimation of source of heavy metal contamination in surface sediments from the Rybink Reservoir. *Chemosphere*, 51, 723-733.

Maata, M., Singh, S. (2008) Heavy metal pollution in Suva harbor sediments, Fiji. *Environment Chemistry Letter*, 6, 113-118.

Marchand, C., Lallier-Vergès, E., Baltzer, F., Albéric, P., Cossa, D., Baillif, P. (2006) Heavy metals distribution in mangrove sediments along the mobile coastline of French Guiana. *Marine Chemistry*, 98, 1-17.

Martin, J.M., Whitfield, M. (1983) The significance of the river input of chemical elements to the ocean. In: Wong, C.S., Boyle, E., Brul, K.W., Burton, J.D., Goldberg, E.D. (Eds.), Trace Metals in Sea Water. Plenum Press, New York, pp. 265–296.

Mishra, V. K., Kim, K. H., Kang, C. H., Choi, K. C. (2004) Wintertime sources and distribution of airborne lead in Korea. *Atmosphere Environment,* 38, 2653–64.

Mora, S., Sheikholeslami, M. R., Wyse, E., Azemard, S., Cassi, R. (2004) An assessment of metal contamination in coastal sediments of the Caspian Sea. *Marine Pollution Bulletin*, 48, 61-77.

Müller, J., Ruppert, H., Muramatsu, J., Schneider, J. (2000) Reservoir sediments – a witen of mining and idjstrialdevel (MalterReservor, estern Erzgebirge, Germany).*Environmental Geology*, 39 (12), 1341-1351.

Nemr, A. E., Khaled, A., Sikaily, A. E. (2006) Distribution and statistical analysis of leachable and total heavy metals in the sediments of the Suez Gulf. *Environmental Monitoring and Assessment*, 118, 89-112.

Nolting, R. F., Ramkema, A., Everaarts, J. M. (1999) The geochemistry of Cu, Cd, Zn, Ni and Pb in sediment cores from the continental slope of the Banc d'Arguin (Mauritania). *Continental Shelf Research*, 19, 665-691.

Onwueme, V., Feng, H.(2006)Risk characterization of contaminants in Passaic River sediments, New Jersey.*Middle States Geographer*, 39, 13-25.

Pazi, I. (2010) Assessment of heavy metal contamination in Candarli Gulf sediment, Easter Aegean Sea. *Environmental Monitoring Assessment*, 174, 199-208.

Quevauviller, P., Lavigne, R., Cortez, L. (1989) Impact of industrial and mine drainage wastes on the heavy metal distribution in the drainage basin and estuary of the Sado River (Portugal). *Environmental Pollution*, 59, 267–86.

Reddy, M.S., Basha, S., Kumar, V. G. S., Joshi, H. V., Ramachandraiah, G. (2004) Distribution, enrichment and accumulation of heavy metals in coastal sediments of Alang-Sosiya ship scrapping yard, India. *Marine Pollution Bulletin*, 48, 1055-1059.

Reimann, C., de Caritat, P. (2000) Intrinsic flaws of element enrichment factors (EFs) in environmental geochemistry. *Environmental Science Technology*, 34, 5084– 5091.

Salomons, W. 1995. Environmental impact of metals derived from mining activities: Processes, predictions, prevention. *Journal of Geochemical Exploration*, 52, 5-23.

Salomons, W., Forstner, U. (1984) Metals in the Hydrocycle. Springer-Verlag, Berlin. 349 pp.

Santos, I. R., Silva-Filho, E. V., Schaefer, C. E., Albuquerque-Filho, M. R., Campos, L. S. (2005) Heavy metal contamination in coastal sediments and soil near the Brazilian Antarctic Station, King George Island. *Marine Pollution Bulletin*, 50, 185-194.

Schropp, S. J., Lewis, F. G., Windom, H. L., Ryan, J. D., Calder, F. D., Burney, L. C. (1990) Interpretation of metal concentrations in estuarine sediments of Florida using aluminum as a reference element. *Estuaries*, 13, 227-235.

Shumilin, E. N., Carriquiry, J. D., Camacho-Ibar, V. F., Sapozhnikov, D., Kalmykov, S., Sánchez, A., Aguíñga-García, S., Sapozhnikov, Y. A. (2002) Spatial and vertical distributions of elements in sediments of the Colorado River delta and Upper Gulf of California. *Marine Chemistry*, 79, 113-131.

Sinex, S. A., Helz, G. R. (1981) Regional *geochemistry of trace elements in Chesapeake Bay sediments. Environmental* Geology, 3, 315-323.

Sinex, S. A., Wright, D. A. (1988) Distribution of trace metals in the sediments and biota of Chesapeake Bay.*Marine Pollute Bulletin*, 19, 425-431.

Spagnoli, F., Bartholini, G., Dinelli, E., Giordano, P. (2008) Geochemistry and particle size of surface sediments of Gulf of Manfredonia (Southern Adriatic sea). *Estuarine, Coastal and Shelf Science*, 80, 21-30.

Sundaray, S. K., Nayak, B. B., Lin, S., Bhatta, D. (2011) Geochemical speciation and risk assessment of heavy metals in the river estuarine sediments-A case study: Mahanadi basin, India.*Journal of Hazardous Materials*, doi:10.1016/j.jhazmat.2010.12.081

Sutherland, R. A., Tolosa, C. A., Tack, F. M. G., Verloo, M. G. (2000) Characterization of selected element concentrations and enrichment ratios in background and anthropogenically impacted roadside areas. *Environmental Contamination Toxicology*, 38, 428-438.

Taylor, S. R., McLennan, S. M. (1995) The geochemical evolution of the continental crust. *Reviews of Geophysics*, 33, 241-265.

Taylor, S. R., McLennan, S. M. (1995) The geochemical evolution of the continental crust. *Reviews of Geophysics*, 33, 241-265.

Topcuoğlu, S., Kirbaşoğlu, Ç., Yilmaz, Y. Z. (2004) Heavy metal levels in biota and sediments in the northern coast of the Marmara sea. *Environmental Monitoring and Assessment*, 96, 183-193.

Trefry, J. H., Shokes, R. F. (1981) History of heavy metal inputs to Mississippi Delta sediments. *Marine Environmental Pollution*, 2, 193-208.

Underwood, E. J. (1971) Trace Elements in Human and Animal Nutrition (3rd Edition).Academic Press, Inc., New York.

Urban, D. L., Cook, N. J. (1986) Hazard Evaluation, Standard Evaluation Procedure, Ecological Risk Assessment. U.S. Environmental Protection Agency, Washington, D.C. EPA540-9-85-001.

Valette-Silver, N. J. (1993) The use of sediment cores to reconstruct historical trends in contamination of estuarine and coastal sediments. *Estuaries*, 16, 577-588.

Villaescusa-Celaya, J. A., Gutiérrez-Galindo, E. A., Flores-Muňoz, G. (2000) Heavy metals in the fine fraction of coastal sediments from Baja California (Mexico) and California (USA). *Environmental Pollution*, 108, 453-462.

Wang, X. L., Tao, S., Dawson, R. W., Xu, F. L. (2002) Characterizing and comparing risks of polycyclic aromatic hydrocarbons in a Tianjin wastewater-irrigated area.*Environmental Research,* 90, 201-206.

Windom, H. L., Schropp, S. J., Calder, F. D., Ryan, J. D., Smith Jr., R. G., Burney, L. C., Lewis, F. G., Rawlinson, C. H. (1989). Natural trace metal concentrations in estuarine and coastal marine sediments of the southeastern United States.*Environmental Science Technology*, 23, 314-320.

Windom, H., Byrd, J., Smith, R., Hungspreugs, M., Dharmvanij, S., Thumtrakul, W., Yeats, P. (1991) Trace metal-nutrient relationships inestuaries.*Marine Chemistry*, 32, 177-194.

Yongming, H., Peixuan, D., Junji, C., Posmentier, E. S. (2006) Multivariate analysis of heavy metal contamination in urban dusts of Xi'an. Cent. China. *Science Total Environment*, 355, 176-186.

Yu, K. C., Tsai, L. J., Chen, S. H., Ho, S. T. (2001) Correlation Analyses on Binding Behavior of Heavy Metals with Sediment Matrices. *Water Resource*, 35, 2471-2428.

Zhang, J., Liu, C. L. (2002) Riverine composition and estuarine geochemistry of particulate metals in China-weathering feartures, anthropogenic impact and chemical fluxes.*Estuarine Coastal Shelf Science*, 54, 1051-1070.

Zhang, L., Ye, X., Feng, H., Jing, Y., Ouyang, T., Yu, X., Liang, R. and Chen, W. (2007) Heavy metal contamination in western Xiamen Bay sediments and its vicinity, China.*Marine Pollution Bulletin*, 54, 974-982.

Zhang, W., Feng, H., Chang, J., Qu, J.,Xie, H., Yu, L. (2009) Heavy metal contamination in surface sediments of Yangtze River intertidal zone: an assessment from different indexes . *Environmental Pollution*, 157, 1533–1543.

Zourarah, B., Maanan, M., Carruesco, C., Aajjane, A., Mehdi, K., Freitas, M. C. (2007) Fifty-year sedimentary record of heavy metal pollution in the lagoon of Oualidia (Moroccan Atlantic coast). *Estuary, Coastal and Shelf Science,* 72, 359-369.

In: Heavy Metal Sediments
Editor: Lucía H. Sanz

ISBN: 978-1-61470-690-8
© 2012 Nova Science Publishers, Inc.

Chapter 4

Heavy Metals and Iron Contamination of Superficial Sediments in the Bizerte Lagoon, Northern Tunisia

*Amor Ben Garali**
Faculté des Sciences de Bizerte,
Laboratoire de sédimentologie et géochimie marine,
Zarzouna, Bizerte, Tunisie

Introduction

The depression occupied by Bizerte lagoon (northeast of Tunisia) corresponds to ante-Neogene solid mass ploughed up and eroded from Messinian (Burollet 1951; Andrieu 1971). This area presents several wadis, in particular wadi Tinja (WT), natural gathered of the garaet Ichkeul, which is considered as the biggest surface of fresh water in Tunisia, and it is in permanent connection with the sea (Figure 1). The shoreline lagoon is urbanized and industrialized (e.g. steelworks of Menzel Bourguiba).

Many works (Salomons 1980; Forstner and Whittmann 1981; Span 1993; Azzaoui 1993; Added 2002; Meador et al. 2005) show clearly the important role played by sediments in the pollutants geochemical cycle. Soussi (1981)

* Tel : 216 97 135 055 - Fax : 216 72 590 566. E-mail: bgamor@yahoo.fr.

indicates that the Bizerte lagoon sediments have primarily a muddy nature to a sandy-muddy containing organic matter.

Watkins and Pilkey (1973), show that heavy minerals present less than 1% of the sandy sediment of Bizerte bay. Yoshida (2004) supposes that one of the metallic pollution sources is the industrial activities, particularly in the vicinity of Menzel Bourguiba town; the other one is probably caused by pesticides pollution in the neighbouring agricultural zones and finally the urban rejections. (Srarfi, 2007) indicates that the maximum of organic pollution is recorded in the south-western part of the Bizerte lagoon. The mixture zones of fresh and salty waters constitute an important environment of marine chemical reactions, which are consecutive to the sudden change of environmental conditions (Ouakad 2007). This type of environment corresponds to estuary, delta or coastal lagoon with continental and marine influences.

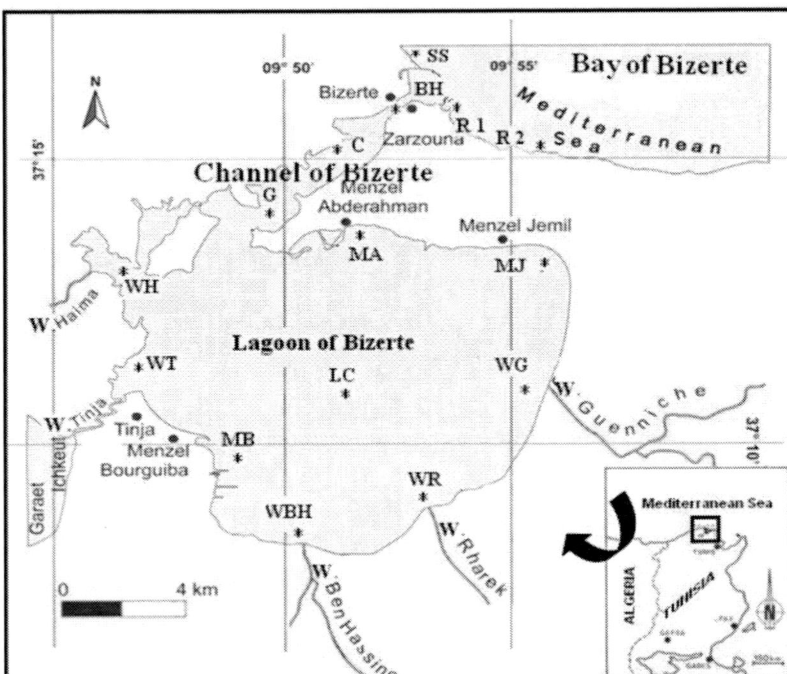

SS: Sidi Salem, R1: R'mel 1, R2: R'mel 2, BH: Before Harbour, C: Channel, G: Gully, WH: Wadi Haima, WT: Wadi Tinja, MB: Menzel Bourguiba, WBH: Wadi Ben Hassine,

WR : Wadi Rharek, WG : Wadi Guenniche, MJ: Menzel Jemil, MA: Menzel Abderhman et LC: Lagoon Centre.

Figure 1. Sampling locations on the Bizerte lagoon.

These industrial activities are the main sources of pollution for this fragile ecosystem, particularly by heavy metals and iron. Fine particles, clays and organic matter are in fact able to trap and to concentrate the heavy metals by adsorption and complexation mechanisms). The sediment metallic load can have a mobilization possibility. In this context, water contamination is the origin of migration, by desorption, salting out and dissolution of heavy metals from the sediment towards water (Jain 2004). The impact study of the different urban and industrial rejections poured in Bizerte lagoon consists in this work analyzing the contamination degree of the surface sediments by heavy metals and iron, as well as a possible organic pollution, in order to safeguard this fragile lagoon ecosystem.

1. Lithology and Clay Cortege of the Superficial Sediments

The superficial sediments of the Bizerte lagoon show a seasonal variation of lithologic features by sector. During summer and autumn seasons, deposits at wadi mouths are muddy, silty and sandy, with 64% of fine fraction. During winter season from January to May, the superficial sediments are represented by fine grained fraction (clays, silts and even mud, until 98% of fine fraction; Ben Garali et al. 2008). This clayey fraction comes from erosion of recent lands (marls and silt) and filed by wadis into the lagoon. These results show the importance of the fine fraction in the superficial sediments of the Bizerte lagoon during all the year. The kaolinite clay is more abundant during all seasons (21-68%), followed by the illite with 14-46%, then smectite (11-35%), and finally chlorite (4-15%; Ben Garali et al. 2008).

2. Total Organic Carbon (TOC) and Total Nitrogen (Nt)

The superficial sediments of the lagoon and the bay of Bizerte contents in organic total carbon do not show any significant temporal fluctuation. The superficial sediments of the bay of Bizerte are less rich in total organic carbon (TOC; 0.18-1.21%) than those of the lagoon (0.49-3.45 %), during this survey period.

The superficial sediments of Menzel Abderhman (MA) and Menzel Bourguiba (MB) localities show the highest contents of TOC respectively with 3.45% and 2.8%. The high levels recorded are related to the urban and industrial worn water rejections which are rich in organic matter. The enrichment in organic carbon at the deltaic zones was observed by several authors in similar environments, particularly, the Rhone delta (France, Aloisi et al. 1975) and Garaet Ichkeul (Tunisia, Ouakad 1982). At the Tinja delta (WT) and Haima delta (WH) stations, we observe that the high contents in organic carbon are associated with smectite enriched sediment. In this prodeltaic zone, we observe an organ-mineral and/or electrochemical flocculation phenomenon. In the bay of Bizerte, the superficial sediments show a low TOC content (below 1%); this can be explained by the comparatively high contents of the sediment in coarse fraction, knowing that there is an incompatibility between coarse sediments and increases of the TOC contents (Gadel 1975).

The total nitrogen contents at the superficial sediments of the Bizerte lagoon, as well as those of the bay of Bizerte, oscillate between 0.7% and 2.1%. The spatial fluctuation map indicates two concentric rings at MA and wadi Ben Hassine (WBH) stations. Yoshida (2004) show an organic important concentration at the Bizerte lagoon because of the alluvial provisions of WT and wadi Haima (WH). The same author indicated an organic origin of the high sedimentary content in total nitrogen at the station of MA.

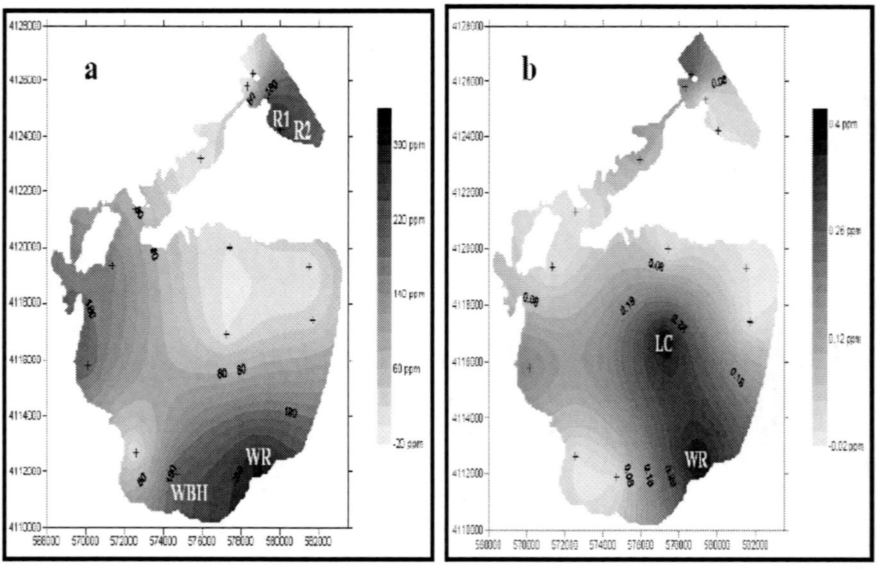

Figure 2. Spatial distribution of the lead (a) and the cadmium (b) in the superficial sediments of the Bizerte lagoon.

3. The Heavy Metals

3.1. Lead

The levels of lead are often very high at the stations of the WR (303.9 ppm), R'mels (R2, 249.8 ppm and R1, 235.2 ppm) and WBH (215.8 ppm; Figure 2a). Ouakad (2007) mentioned that the high levels of lead in the superficial sediments are oriented along an axis NE-SW, which corresponds probably to fossil residues of lead ore washing products from the old treatment factory, situated on the north coastal of the lagoon.

According to US NOAA (1999) norms with 300 ppm, only wadi Rharek station (WR) with 303.9 ppm can be considered lightly contaminated by lead. The other stations are far of all significant contamination by the lead. Nevertheless, according to the reference value of 30 ppm adopted by RNO (1997), all the stations of the Bizerte lagoon are considered contaminated, since the lowest value is of 32.1 ppm, in wadi Guenniche (WG) (Figure 2a).

3.2. Cadmium

Cadmium is the less widespread metal in the superficial sediments of the lagoon and the bay of Bizerte; it is detected only at the stations of the wadi R'harek (WR) and at the lagoon centre with a content of 0.395 ppm (Figure 2b). At the stations of Sidi Salem (SS), WT, the zone before harbour and channel (C), with contents respectively of 0.26, 0.195, 0.16 and 0.11 ppm (Figure 2b). Yoshida (2004) mentioned very close contents that mine at the sediments of surface of the Bizerte lagoon (0.04-0.53 ppm). The low cadmium content mentioned by Hedfi (2003) in the superficial sediments of the old harbour of Bizerte are very close to our results, namely, 0.78-1.51 ppm. According to the norms adopted by the RNO (1997) and the US NOAA (1999), we noted the absence of any significant contamination by the cadmium, since the sedimentary contents taken are below to 2 and 3 ppm.

3.3. Zinc

The zinc is the more widespread metal at the superficial sediments both in the lagoon and in the bay of Bizerte. The strongest Zinc content (749.8 ppm) is recorded in front of Menzel Bourguiba station (MB), while the lowest content is recorded just beside, at the WT station, with 32 ppm (Figure 3a). In the bay, the sandy sediments are judged rich in zinc, with contents between 120.4 and 204.6 ppm (Figure 3a). These results are in accordance with the "Office de la marine marchande et des ports" (OMMP 2000) results, which indicate that the high zinc contents (of 204-226 ppm) are recorded at the superficial sediments of MB station. According to the admissible value by the RNO (1997), with 90 ppm, only WT station is considered not contaminated by zinc. Nevertheless, according to US NOAA (1999) norms (with 720 ppm), only the superficial sediments of the MB station are revealed contaminated.

3.4. Copper

The superficial sediments of WH prodeltaic zone show the highest contents of copper with 67.4 ppm. The copper is also present but with less important contents at WBH station (42.6 ppm). The spatial fluctuation map indicates concentric rings at these stations (Figure 3b). Nevertheless, the lowest content was recorded at MB station, with 1 ppm. The contents suggested by Cheikh (2003) in the same environment, are close to our results, namely the values oscillate between 1.67 and 25.24 ppm. Srarfi (2007) shows a concentration of metallic elements (Pb, Cd, Zn, and Cu) at the south-western part of the Bizerte lagoon, which is caused by the waste disposal of Menzel Bourguiba city. The sedimentary contents in copper did not surpass the admissible limit by US NOAA (1999), which is of 390 ppm, while the sedimentary contents of WH and WBH stations are, respectively, 67.4 and 42.6 ppm and are beyond the admissible limit by the RNO (1997) with 30 ppm, and therefore, they are considered contaminated.

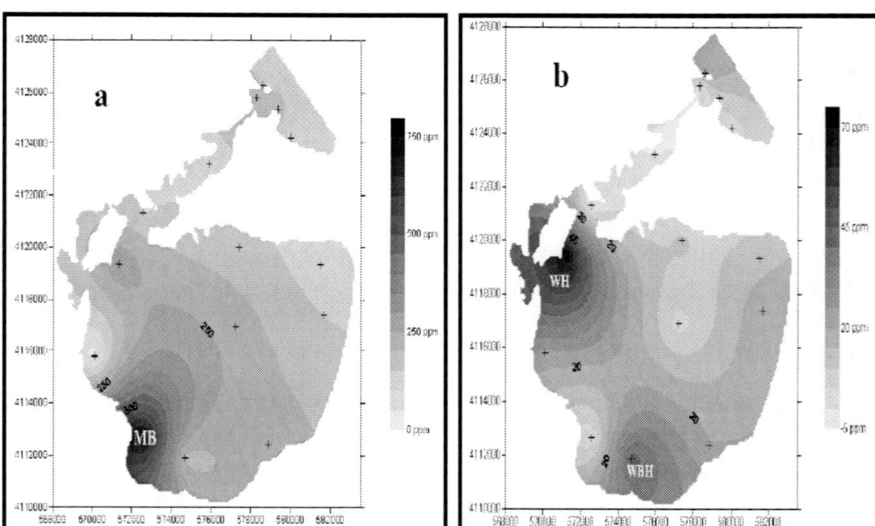

Figure 3. Spatial distribution of the zinc (a) and the copper (b) in the superficial sediments of the Bizerte lagoon.

3.5. Nickel

The nickel is very widespread at the superficial sediments of Menzel Jemil and Gully (G) stations, with respectively, 192 and 122.9 ppm (Figure 4a). As for zinc, nickel is very widespread also at the superficial sediments of MB station, with 119.5 ppm (Figure 4a). These results are compatible with those given by the (OMMP, 2000), where the maximum of nickel is recorded at the channel (C) and MB stations, respectively, 60 and 65 ppm. The sedimentary contents in nickel at the Bizerte lagoon mentioned by Yoshida (2004) are very low in comparison with our results, namely between 10 and 25 ppm. According to the admissible value by the US NOAA (1999), which is of 110 ppm, only the sediments of Menzel Jemil, Gully and MB stations are considered contaminated by the nickel.

3.6. The Manganese

The spatial distribution map of the manganese contents in the superficial sediments of Bizerte lagoon shows a principal concentric ring at WT mouth station, with a content of 178.7 ppm, and a second concentric ring at WH mouth station, with a content of 107.5 ppm (Figure 4b). Essid (1999), showed close values to our results, namely 51.9-104.3 ppm.

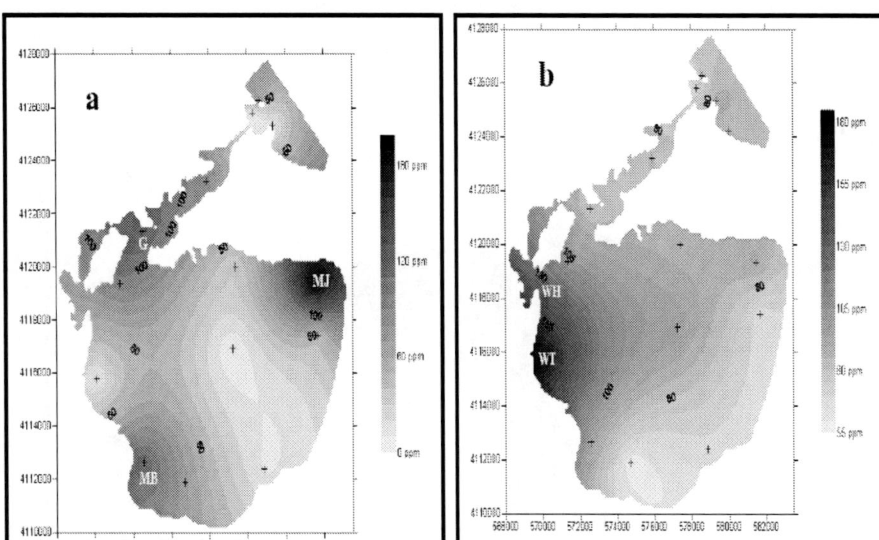

Figure 4. Spatial distribution of the nickel (a) and (b) in the superficial sediments of the Bizerte lagoon.

The recorded sedimentary contents in manganese are inferior to 260 ppm [value admissible by the US NOAA (1999)] and do not reveal any significant contamination of the sediments by this metal.

4. Iron

An alkaline pH, the iron precipitates very quickly in the form of an oxide (haematite Fe_2O_3) and hydroxide (goethite FeO OH). The contents of Bizerte lagoon superficial sediments in iron oscillate between 0.39% at SS station and 3.76% at MB station. This enrichment is probably related to the steelworks rejections of Menzel Bourguiba city (Figure 5). The WT, MA, Menzel Jemil and WBH stations show also high contents in iron, respectively, of 2.22%; 1.81%; 1.67% and 1.24 %. Srarfi (2007) suggested high values in iron in the superficial sediments of the Bizerte lagoon between 0.5% and 5%. According to Izquierdo et al. (1997), sediments containing average concentrations in iron of 0.75% can be considered as not contaminated. It means that the Bizerte lagoon surface sediments are considered not very polluted in iron. These values did not surpass the norm of 22% adopted by the US NOAA (1999).

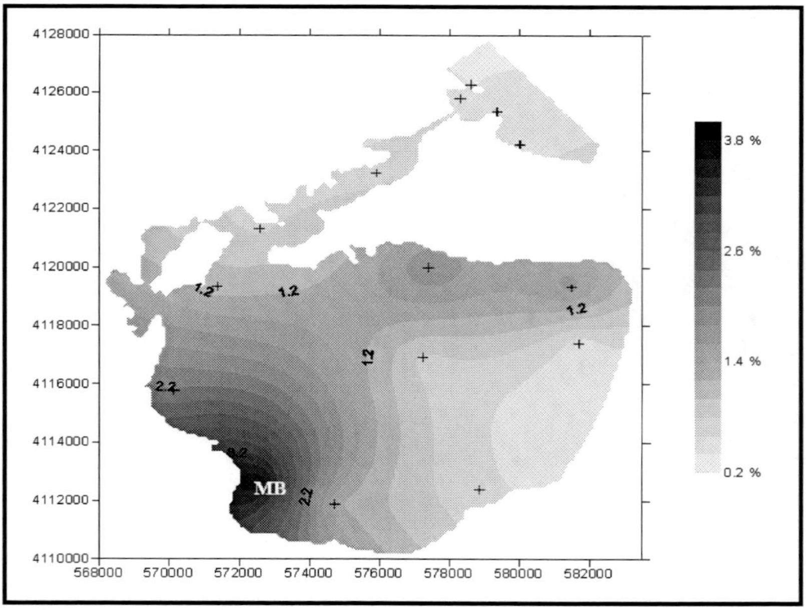

Figure 5. Spatial distribution of the iron in the superficial sediments of the Bizerte lagoon.

Table 1. Comparison of the average contents of certain heavy metals of the superficial sediments of the Bizerte lagoon

Authors / Metals	Kamens and al., (1989)	Essid (1999)	Dellali (2001)	Ouakad (2007)	Srarfi (2007)	Present work
Mn (ppm)	15-102	100-117	50-127	459	50-650	31,2-178,7
Zn (ppm)	6-153	58-256	128-181	247	50-1700	32-749,8
Cu (ppm)	0,5-10,5	10-25	45-67	58	5-140	1-67,4
Pb (ppm)	12-52	46-135	103-117	68	20-320	32,1-303,9
Ni (ppm)	-	47-105	19-45	-	4-24	2,6-192

The comparison of our results with those of the literature (Kamens et al. 1989; Essid 1999; Dellali 2001; Srarfi 2007 and Ouakad 2007) shows a remarkable increase of the concentrations in various heavy metals in the Bizerte lagoon (Table 1).

5. Interpretation

The north-western sector of the lagoon shows a contamination of the superficial sediments by nickel, probably as a consequence of the neighbouring industrial zone. The south-eastern sector shows a contamination by lead, result of a detrital material enriched in lead from Jbel Kchabta and Jbel Messeftine geological lands. At the south and eastern sectors, the contamination of the superficial sediments is essentially by copper, zinc, nickel and iron.

This pollution is related to the rejections of the steelworks, military arsenal and Menzel Bourguiba waste disposal (Figure 6). In the interconnexion zone of the lagoon with the gully, the pollution of the superficial sediments is principally by nickel (Figure 6). The spatial evolution of the metals concentrations seems to nearly follow the sedimentary percentages in fine particles. The normalization factors show that the majority of elements have strong affinity for the phyllosilicates and organic matter.

In the present work, the metallic types show a positive correlation, respectively, with the fine fraction and the total nitrogen, such as Pb (0.23 and 0.20), Cd (0.13 and 0.48), Ni (0.19 and 0.20), Cu (0.20 and 0.22) and the Fe (0.52 and 0.62; threshold 5%, Table 2). Ouakad (2007) indicates that the higher contents in copper are situated at the central sector sediments which are clayey features and organic matter enriched. The Zn (0.13 and 0.11) and the

Mn (0.25 and 0.48; threshold 5%) are correlated positively, respectively with fine fraction and TOC (Table 2).

We notice a preferential association between smectite and TOC ($r = 0.60$; threshold 5%, Table 2). Zn tends to associate preferentially with the matrix clayey by adsorption and with the organic matter by chelating. Ouakad (2007) shows that the zinc high contents of the sediments characterize the deltaic zone of wadi Tinja (WT) and the central sector (with clayey features). The same author noted that the sedimentary content in Zn and Mn tends to decrease in the coastal zones (with sandy features), which are conformed to our results. The same phenomenon was observed in the Garaet Ichkeul and in Ghar el Melh lagoon by Ouakad (1982 and 2007). A good correlation is noted between clayey phase and manganese high contents (0.23, 0.22 and 0.21, respectively for the illite, kaolinite and smectite, threshold 5%, Table 2). The zones enriched by manganese and zinc correspond to the higher contents of kaolinite and smectite (Ben Garali and al. 2008).

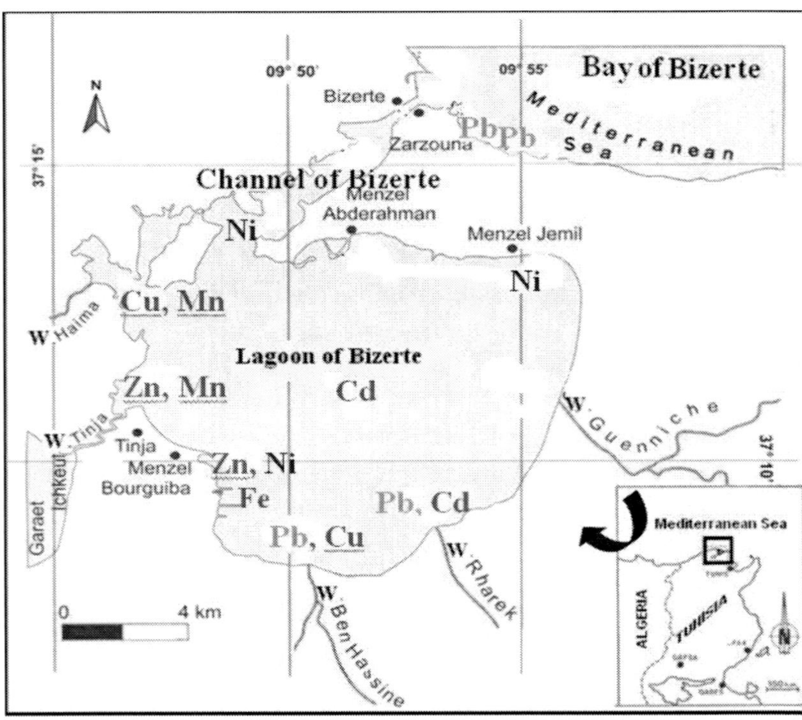

Figure 6. Spatial distribution of the heavy metals and iron in the superficial sediments of the Bizerte lagoon.

Table 2. Correlations between different analysed parameters from the superficial sediments of Bizerte lagoon

	FF	K	S	I	C	TOC	Nt	Pb	Cd	Zn	Ni	Cu	Mn	Fe
FF	1													
K	0,95	1												
S	0,96	0,98	1											
I	0,96	0,99	0,99	1										
C	0,91	0,97	0,95	0,96	1,00									
TOC	0,71	0,65	0,60	0,63	0,60	1								
Nt	0,03	-0,10	-0,07	-0,09	-0,01	-0,09	1							
Pb	0,24	-0,02	0,00	0,01	-0,06	-0,27	0,20	1						
Cd	0,14	-0,07	0,03	-0,01	-0,01	-0,04	0,49	-0,16	1					
Zn	0,14	0,08	0,10	0,09	-0,02	0,11	-0,30	-0,27	-0,11	1				
Ni	0,20	0,36	0,29	0,31	0,33	0,21	0,20	-0,08	-0,52	0,09	1			
Cu	0,20	0,18	0,21	0,23	0,18	-0,19	0,22	0,42	0,01	-0,18	0,07	1		
Mn	0,25	0,22	0,21	0,23	0,20	0,49	0,20	0,04	0,10	-0,39	-0,12	0,10	1	
Fe	0,53	0,53	0,56	0,55	0,44	0,63	0,63	-0,19	-0,26	0,19	0,27	0,09	0,50	1

FF: Fine Fraction, K: Kaolinite, S: Smectite, I: Illite, C: Chlorite, TOC: Total Organic Carbon, Nt: total Nitrogen, Pb: Lead, Cd: Cadmium, Zn: Zinc, Cu : Copper, Ni : Nickel, Mn: Manganese, Fe: Iron.

Therefore, there would be an association with the finest phyllosilicates and/or with the colloidal oxyhydroxydes. The strongest correlation between the metallic types is between the manganese and iron ($r = 0.50$; threshold 5%, Table 2). Iron and manganese of anthropogenic origin, present in the lagoon are principally originated by industrial rejections and, in a least measure by urban and agricultural rejections (Ben Guirat 1997).

CONCLUSION

According to the norms adopted by the RNO (1997) and the US NOAA (1999), the majority of superficial sediments of Bizerte lagoon is judged contaminated by lead, zinc and nickel and at a least degree by copper and iron. The very significant positive correlations recorded between the fine fraction and TOC sedimentary contents and heavy metals, explain that the fine sediment is generally endowed with a high adsorption capacity than crude sediment of same origin; therefore it is more able to trap the organic matter and to set up other contaminations as metals (Zourarah et al. 2001).

The sedimentary contents variations observed in metallic elements can be explained by an anthropogenic provision of the metallic elements and/or fluctuations of the environment physic-chemical parameters during the time. The analysis results of the superficial sediments show that the majority of the analyzed elements are "argilophilous" except the zinc and the manganese which are clayey organophile.

REFERENCES

Added A (2002) Cycles biogéochimiques des sels nutritifs, du fer, du manganèse et du soufre dans les sédiments de deux systèmes côtiers du nord de la Tunisie: lagune de Ghar el Melh et du lac nord de Tunis. Thèse ès-Sciences. Université de Tunis el Manar, Tunis, Tunisie, p 266.

Aloisi JC, Barusseau JP, Monaco A (1975) Rôle des facteurs hydro-dynamiques dans la sédimentation sur les plateaux continentaux. Comptes Rendus de l'Académie des Sciences. Paris, 280 (série D): 579-582.

Andrieu M (1971) Prospections électriques au sud du lac de Bizerte. Rapport number 23: 10 p. Direction des Ressources et Sciences de l'Eau. Ministre de l'agriculture, Tunis, Tunisie.

Azzaoui S (1993) Géochimie de certains métaux lourds dans les sédiments de la lagune de Ghar el Melh. DEA. Univ. De Tunis II. Fac. Sc. Tunis. 122 p.

Ben Garali A, Ouakad M, Gueddari M (2008) Evolution spatio-temporelle du cortège argileux dans les sédiments superficiels de la lagune de Bizerte (Tunisie septentrionale). *Afr. Geosci. Rev.* 15 (3): 279-286.

Ben Guirat S (1997) Contribution à l'étude de la pollution des terres agricoles par des rejets industriels et miniers. (Cas de la sidérurgie de Menzel Bourguiba et de la mine de Bougrine). DEA, Université de Tunis II, Tunis, Tunisie, p 127.

Burollet PF (1951) Etude géologique des bassins mio-pliocènes du Nord - Est de la Tunisie. Annales de Mines et Géologie, Tunis, 1ére série : Géologie générale et régionale, number 7, pp5 – 86.

Cheikh M (2003) Détermination des pesticides organochlorés et des métaux lourds dans les sédiments de la lagune de Bizerte. D.E.A. Sci. Env. F.S.B. 76 p.

Dellali M (2001) Utilisation d'indicateurs microbiologiques et biochimiques chez Ruditapes Decussatus et Mytilus galloprovincialis dans la biosurveillance de la lagune de Bizerte : Validation de certains biomarqueurs. Thèse de Doctorat. Université 7 Novovembre. à Carthage, Tunis, Tunisie, p 215.

Essid N (1999) Contribution à l'étude de la pollution organo- minérale de la lagune de Bizerte. Caractérisation géochimique et sédimentologique et impact sur les nématodes libres marins. DEA. Université 7 Novovembre à Carthage, Tunis, Tunisie, p 189.

Forstner U, Wittmann GTW (1981) Metal pollution in the aquatic environnement. Springer-verlag, Berlin, p 486.

Gadel F (1975) La matière organique et la sédimentation récente dans le golfe du lion. Distribution, nature et évolution. Thèse d'état. Université de Paris, France, p 263.

Hedfi A (2003) Etat de pollution du vieux port de Bizerte et impact sur le meiobenthos. DEA. Université. 7 Novovembre à Carthage, Tunis, Tunisie, p 91.

Izquierdo C, Usero J, Garcia I (1997) Speciation of heavy metals in sediments from salt marshes on the southern atlantic coast of spain. *Mar. Poll. Bull.* 34: 123-128.

Jain CK (2004) Metal fractionation study on bed sediments of rivers Yamuna, *India. Water Res.*, 38: 569-578.

Kamens J, Pilkey O, Whaling P, Harbridge W Chaouch A (1989) Etude sédimentologique des lacs de Bizerte et de Tunis. Notes du Service Géologique. Office National des Mines, number 57, Tunis, Tunisie.

Meador JP, Ernest DW, Kagley AN (2005) A comparison of the non-essential elements cadmium, mercury, and lead found in fish and sediment from Alaska and California. *Science of the Total Environment*, 339:189-205.

Ommp (2000) Office de la Marine Marchande et des Ports. Etude d'impact sur l'environnement des travaux de dragage aux ports de Sousse et de Bizerte/Menzel Bourguiba. Rapport préliminaire, number1, Tunis, Tunisie, 110 p.

Ouakad M (2007) Genèse et évolution des milieux laguno-lacustres du nord-est de la Tunisie (Garaet Ichkeul, Lagunes de Bizerte et de Ghar el Melh). Thèse ès. Sciences Géologiques. Université de Tunis el Manar. Tunis, Tunisie, 453 p.

Ouakad M (1982) Etude sédimentologique et caractères géochimiques des dépôts récents de la Garaet el Ichkeul (Tunisie septentrionale). Thèse de 3ème cycle. Université de Perpignan, France, 166 p.

RNO (1997) Tendances temporelles des teneurs en contaminants dans les mollusques du littoral français. Surveillance du Milieu Marin. Travaux du Réseau National d'Observation de la qualité du milieu marin (RNO). Édition 2000. Ifremer et Ministère de l'Aménagement du territoire et de l'Environnement. 9-32. ISSN 1620-1124.

Salomons W (1980) Trace metal analysis on polluted sediments Part II, Evaluation of environmental impact. *Environ. technol. lett.* 1: 506-517.

Soussi N (1981) Mécanismes de la sédimentation et évolution paléogéographique de la lagune de Bizerte (Tunisie) durant le quaternaire récent. Thèse 3 cycle de géologie structurale, Université Paul Sabatier, Toulouse, France, p 229.

Span D (1993) Etude par fractionnement chimique de la distribution des métaux traces dans les sédiments marins récents : le delta du Rhône, les golfes de St Raphael. Doctorat troisième cycle,. Université de Pierre et Marie Curie (Paris VI), France, p 177.

Srarfi F (2007) Etude géochimique et état de pollution de la lagune de Bizerte. Thèse de doctorat, Univ. Tunis el Manar. Tunis, Tunisie, p 122.

US NOAA (1999) Screening Quick Reference Tables (SQuiRTs). Coastal protection and restoration division. National Marine Fisheries Service (NMFS). Our Living Oceans. Report on the status of U.S. living marine resources, 1999. U.S. Dep. Commer. National Oceanic and Atmospheric Administration (NOAA) Tech. Memo.

Watkins J, Pilkey OH (1973) Le sable des plages Tunisienne. Notes Services Géologiques. Office National des Mines. Tunis, Tunisie, number 30: 337-392.

Yoshida M (2004) Mobility of potentially toxic element from the bottom sediment of Bizerte lagoon-sequential leaches and acid extraction. Shipoard Scientists Team. Final report research promotion programmes,. Tunis, Tunisie, pp 55-70.

Zourarah B, Carruesco C, Labraimi M, Reboullon P, Bakkas S (2001) Impact des effluents anthropiques sur les teneurs en métaux lourds dans les sédiments de la lagune de Oualidia (Maroc). *CIESM* 36: 175-183.

In: Heavy Metal Sediments
Editor: Lucía H. Sanz
ISBN: 978-1-61470-690-8
© 2012 Nova Science Publishers, Inc.

Chapter 5

CORRELATION BETWEEN MILLENNIAL MINING ACTIVITIES AND HEAVY METAL POLLUTION IN TWO SOUTHWESTERN SPANISH ESTUARIES

M. Abad,[1] F. Ruiz,[1] M. Pozo,[2] M. I. Carretero,[3] J. Rodríguez-Vidal,[1] L. M. Cáceres,[1] J. M. Muñoz,[4] M. L. González-Regalado,[1] J. G. Pendón,[5] A. Toscano,[1] J. Borrego,[5] M. I. Prudêncio,[6] and M. I. Dias[6]

[1]Departamento de Geodinámica y Paleontología, Universidad de Huelva, Huelva, Spain
[2]Departamento de Geología y Geoquímica, Universidad Autónoma de Madrid, Madrid, Spain
[3]Departamento de Cristalografía, Mineralogía y Química Agrícola, Universidad de Sevilla, Sevilla, Spain
[4]Departamento de Estadística e Investigación Operativa, Universidad de Sevilla. Sevilla, Spain
[5]Departamento de Geología, Universidad de Huelva, Huelva, Spain
[6]Instituto Tecnologico e Nuclear. Sacavém, Portugal

ABSTRACT

The sedimentological, paleontological and geochemical analyses of two cores collected in the southwestern Spanish coast permits to delimitate the effects of different historical mining periods on the heavy metal contents of estuarine sediments. These effects are stronger in the Huelva estuary, with a first period of low to moderate pollution during Roman times (~2000 yr BP) and very high concentrations of heavy metals derived from recent mining activities (19th-20th centuries). On the contrary, sediments remain unpolluted or present very low metal contents in the external part of the Guadalquivir estuary during the last 2400 years, with absolute concentrations of Zn, Cu or Pb associated clearly to the grain size distribution.

1. INTRODUCTION

Estuaries are zones of complex interaction between fluvial and marine processes that may act as a geochemical trap for heavy metals bonded in the fine-grained sediments. In recent years, an increasing number of investigations have used heavy metals to test the effects of industrial, urban or mining pollution in these coastal areas. For these studies, the vertical distribution of differents elements in sediment cores is compared with radiocarbon datings, changes in faunal assemblages or sedimentation rates (Freitas et al., 1999; Köster et al., 2007).

The aims of this chapter is to delimitate the geochemical evolution of two Atlantic Spanish estuaries during the last millenia. Their studied core records of both heavy metals and facies have permitted to reconstruct the palaeoenvironmental evolution of these areas and to evaluate the metal pollution in the different historical periods of these millenial mining activities.

2. STUDY AREA

The southwestern Spanish coast includes four estuaries (Figure 1: Guadiana, Piedras, Tinto-Odiel and Guadalquivir), partly enclosed by prograding barrier spits. Two of them were selected for this research.

2.1. Huelva Estuary

The Huelva estuary is located on the southern Spanish Atlantic coast, at the junction of the Tinto and Odiel rivers (Figure 1), forming a wide drowned valley system incised into a Neogene detrital basement. The marine sector of this estuary is a tidal zone with several bodies of salt marshes (e.g. Bacuta Island) separated by distributary channels protected from the Atlantic ocean by both barrier islands and spits.

Tidal regime, wave action and fluvial discharge are the most dominant factors controlling the hydrodynamic processes in this estuary. The tidal regime is mesotidal and semi-diurnal, with a low diurnal inequality (Borrego et al., 1993). Dominant waves come from the south-west, associated with the Atlantic circulation regime; and the sediment discharge of the Tinto-Odiel river system varies seasonally reaching a maximum between December and January, when rainfalls are abundant on the catchments.

Most of the physical and chemical parameters of water, i.e. salinity, temperature, dissolved oxygen and pH values, undergo significant seasonal variations (IEO, 1992; AMA, 1993). Thus, during the summer, salinity is nearly constant at $36^{0}/_{00}$, owing to the strong tidal current effect; water temperature is about 26°C near surface water; dissolved oxygen content varies from 4 to 6 mg/l, and the pH values range between 6 and 8. In winter, partial stratification occurs where Tinto and Odiel rivers meet, and consequently the surface layer of the water column is markedly less saline (even $5^{0}/_{00}$) and more acidic (pH = 2-5) than the bottom layer (salinity = $29\text{-}32^{0}/_{00}$ and pH = 5-7).

2.2. Doñana National Park

The Guadalquivir estuary (1800 km^2) includes the Doñana National Park, a UNESCO-MAB Biosphere Reserve. This site is one of the largest wetlands (>50,000 ha) in Europe and represents the last tract of relatively undisturbed marshes in this system. These marshes are protected by two coastal barriers (Doñana and La Algaida).

The hydrodynamic processes of this area are controlled by fluvial regime, wave action and drift currents. The Guadalquivir River has a mean annual water discharge of 164 m^3s^{-1}, whereas the tidal regime is mesotidal and semidiurnal. Dominant waves associated with the Atlantic circulation come from the southwest, whereas the main littoral drift currents transport sand-size sediments from the Portugal coast to the Spanish nearshore and contribute to the growth of sandy barriers and spits.

The first historical description of this area was made by the Roman chronicler Strabo in his work *Geographica* (29-7 BC). He referred to the southern part of the Guadalquivir estuary as an inland lagoon (*Lacus Ligustinus*).

2.3. Mining Activity

A wide sector of Southwestern Spain is occupied by the so-called Iberian Pyrite Belt, one of the most important mining areas in western Europe, with giant orebodies of Cu, Zn and Pb that include precious metals (mainly Ag and Au). During the Phoenician (3000 to 2600 years BP) and Roman (2000 and 1800 years BP) periods, numerous Ag-Au deposits were extensively mined, being posteriorly abandoned by the Visigoths (1600 to 1300 years BP) and Muslims (1300 to 500 years BP) (Blanco Freijero and Rothenberg 1981; Ferrero 1988). The pyrite production peaked between 1875 and 1930 and, at present, only few mines (e.g. Las Cruces) are still working.

In addition, wastes containing high levels of heavy metals were discharged from vast industrial concentrations (alkalines, fertilizers, metallurgics) in the Huelva estuary since 1960. Because of the large scale mining and smelting operations and both acid mine-drainage and industrial waste dumping, the Odiel and Tinto rivers present acidic waters and recent estuarine sediments at the Tinto-Odiel junction contain very high concentration levels of toxic, sulphide-associated heavy metals (Fernández Caliani et al., 1997; Ruiz et al., 2001).

3. Methodology

Two cores (Figure 1) were recovered in the Tinto-Odiel estuary (Core A: Bacuta Island) and the Doñana National Park (Core B: Doñana spit), with a length comprised between 11.5 and 31 m, respectively. A preliminary study was developed in order to distinguish the main facies and their main visual features. In a second step, twenty samples were taken and divided into subsamples for the different analyses, according to the facies distribution.

3.1. Grain Size and Palaeontology

Owing to the predominance of detrital facies, the lithological description has been complemented with grain-size analysis. Grain-size distribution was determined by wet sieving for the coarser fractions (>100 µm). Fractions lesser than 100 µm were analysed by photosedimentation (MicromeriticsR SediGraph 5100 ET). In addition, the palaeontological record was studied, including both macrofauna and microfauna. Results derived from these previous analyses permit to infer the environmental evolution of each core.

3.2. Geochemistry

Analyses of major and trace elements of core A were performed by X-Ray Assay Laboratory, Toronto (Canada). Metal concentrations were determined by X-Ray Fluorescence (Al_2O_3, Fe_2O_3) and ICP Mass Spectrometry (Ag, As, Cu, Pb, Zn).

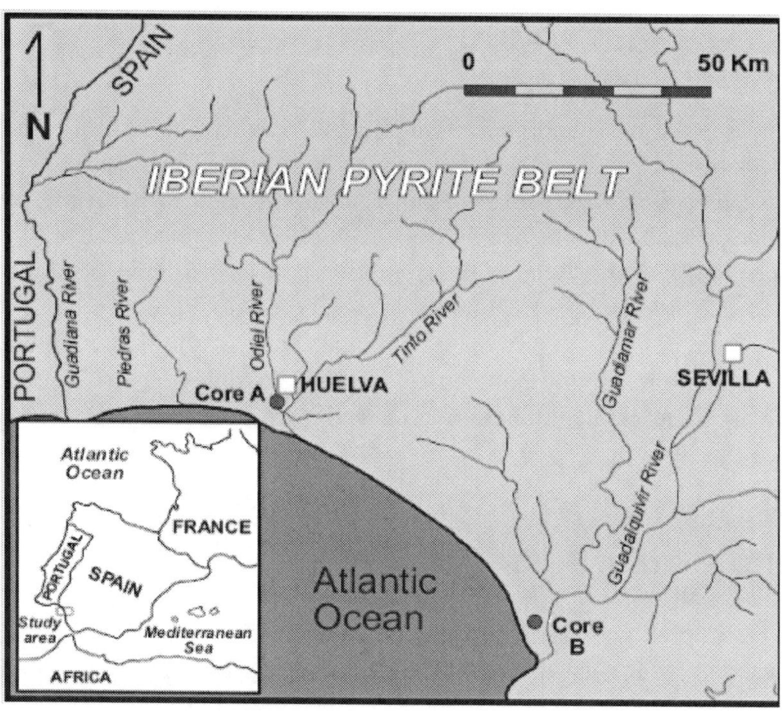

Figure 1. Location of study area and cores.

Chemical analyses for trace elements of core B were performed by Activation Laboratory, Ontario (Canada). Element concentrations were determined by Instrumental Neutron Activation Analysis –INAA- (As), Inductively Coupled Plasma Spectrometry –ICP- (Cu, Pb) or MultiINAA-ICP techniques (Zn).

The calibration is based on the analysis of over 30 international standard reference materials. Detection limits are similar in both laboratories (Al_2O_3: 0.01%; Fe_2O_3: 0.01 %; Ag: 0.2-0.3 ppm; As: 3 ppm; Cu: 0.5-1 ppm; Pb: 2-3 ppm; Zn: 0.5-1 ppm).

3.3. Measure of Metal Pollution

The Index of geoaccumulation (I_{geo}) introduced by Muller (1981) was used as a measure of metal pollution in the sediments:

$$I_{geo} = \log_2 \frac{C_n}{1.5 \times B_n}$$

where C_n is the measured concentration of the element in a sample and B_n is the background of this element in the studied area. The factor 1.5 is used because of possible variations of the background data to lithogenic effects. This background was obtained by Ruiz et al. (1998) for unpolluted silty-clayey sediments of this area (Cu: 13 ppm; Zn: 39 ppm; Pb: 15 ppm). Samples were divided into five groups: unpolluted ($I_{geo}< 1$); very low polluted ($1< I_{geo}< 2$); low polluted ($2< I_{geo}< 3$); moderate polluted ($3< I_{geo}< 4$); highly polluted ($4< I_{geo}< 5$); and very highly polluted ($I_{geo} > 5$). This index has been utilized to assess metal enrichment of both riverine and marine sediments (Borovec et al., 1993; Ruiz, 2001).

3.4. Radiocarbon Dating

One sample was dated at Geochron Laboratory in core A (GX-21287), whereas two additional datings were carried out at Beta Analytic Laboratory (Miami, USA) by radiocarbon methods applied to mollusc shells of core B (B-228876 and B-228873). Data were calibrated using CALIB version 6.0.1, with the regional marine reservoir effect (ΔR) suggested by Soares and Martins (2009) for this area. The final results correspond to calibrated ages (cal.) using 2σ intervals.

4. Results And Discussion

4.1. Brief Description of the Cores: Palaeoenvironmental Reconstruction

The lower 1.5 m of core A consists of well-sorted yellow fine to medium sands. Faunal remains include marine bivalves (*Acanthocardia aculeata*, *Calyptraea chinensis*), gastropods (*Hinia reticulata*), scaphopods and ostracods (*Pontocythere elongata*). These high-energy tidal deposits were deposited in an open estuary scenario (Borrego et al., 1999). These deposits are covered by a chaotic layer with numerous disarticulated and fragmented valves of marine molluscs and marine ostracods that may be attributed to high-energy events (Ruiz et al., 2007). The upper part of this core includes low-energy tidal deposits with an upward decreasing grain size. Foraminifers are abundant (mainly *Ammonia tepida*), whereas ostracods are poorly represented in the upper part of these deposits. Consequently, this core includes a regressive sequence with the progressive infilling of a previous open estuary.

The lower 4 m of core B are constituted by sandy-clayey silts with a mixture of both marine and brackish species, probably deposited near the marine connection of an old lagoon. The mean grain size decreases in the following 8 m, with the presence of numerous brackish species of ostracods such as *Cyprideis torosa* or *Loxoconcha elliptica*, frequent in perimediterranean lagoons (Ruiz et al., 2006). A similar trend was found in the upper part of this core, together with two layers of yellow medium sands with aeolian affinities. The lower sandy horizon represents a high-energy event that eroded the Doñana spit and could be assimilated to the effect of the 218-210 BC tsunami (Campos, 1991). The upper layer includes the recent active dune systems of the Doñana National Park, a UNESCO Biosphere Reserve.

In a general overview, core A represents the progressive infilling of an estuary, whereas the vertical facies distribucion of core B is linked to evolution of an open lagoon, derived from the emersion of the internal areas and the growing of the Doñana spit, with the partial closing of the marine connection. This evolution confirms previous palaeogeographical reconstructions effectuated by Ruiz et al. (2004, 2005).

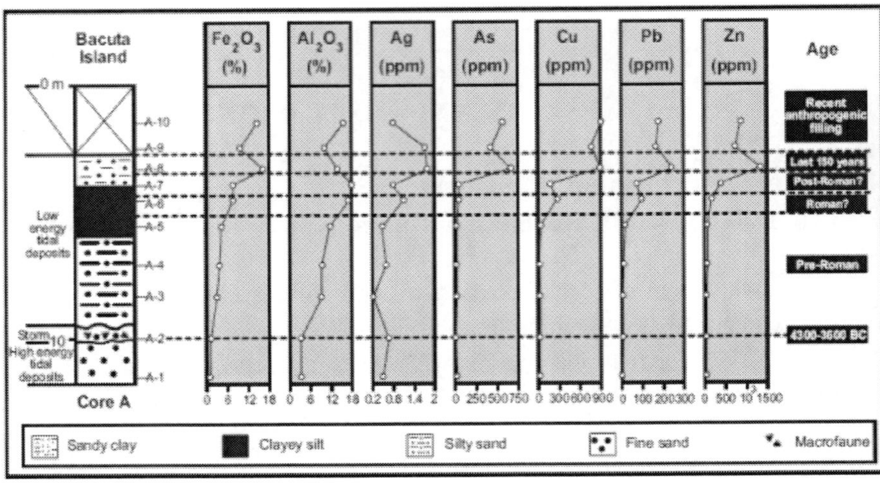

Figure 2. Lithology, geochemistry, palaeoenvironmental evolution and dating of core A.

4.2. Heavy Metal Pollution *vs* Historical Mining

Core A (Huelva estuary). The lower 2 m of core A (sample A-1: high-energy tidal deposits; sample A-2: high-energy layer) are characterized by very low heavy metal contents. This interval was deposited during pre-mining times (> 3600 yr BP). The overlying silty sand and clayey silt (samples A-3 to A-6) shows a progressive increase in Ag, Fe_2O_3 and Al_2O_3 contents, which should be assimilated to the initial explotation of Ag-Au orebodies during the Phoenician phase (3000-2600 yr BP).

Sample A-6 (4.5 m depth) shows a remarkable metal enrichment in relation to the lower samples. This break will indicates the evidence of the first industrial mining by the Romans (~2000 yr BP). Metal contents of sample A-7 are similar or even lower and may be interpreted as a consequence of the interruption in the mining activities during the Islamic domination.

The higher metallic concentrations of core A were determined in sample A-8, taken at 3.3 m depth. This very polluted layer (0.3–0.4 m width) was deposited in the last decades, with high contents of numerous toxic heavy metals derived from recent industrial wastes (1960-1990) and the acidic mining drainage processes.

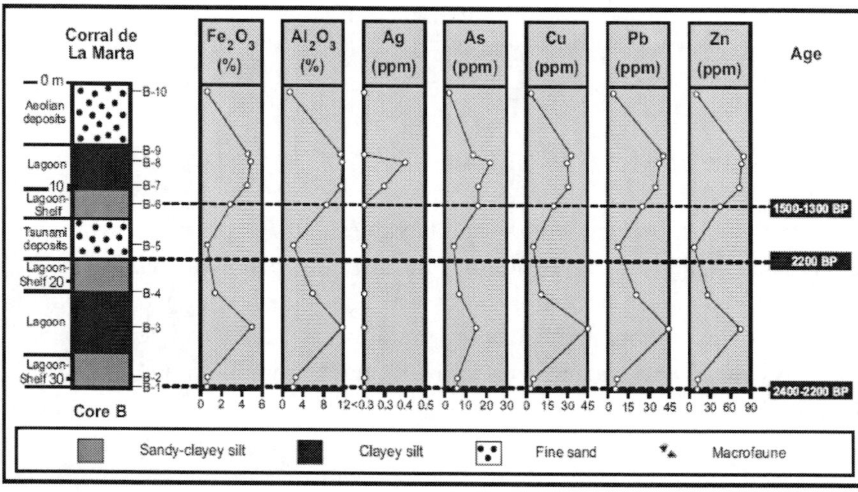

Figure 3. Lithology, geochemistry, palaeoenvironmental evolution and dating of core B.

This layer has been found in other cores collected in the Tinto-Odiel Estuary (Borrego, 1992; González-Regalado et al., 1996).

The upper 3 m (samples A-9 and A-10) belong to an anthropogenic filling with recent surface sediments dredged in the main channel of the Odiel River and wastes from saltworks located north of Bacuta Island. Fernández et al. (1997) obtained similar metallic contents analyzing the bulk metal content of the subtidal sediments of the estuary mouth.

Core B (Doñana National Park). In this core, it is very difficult to distinguish the historical pollution derived from the different mining phases, owing to three different factors: a) a different palaeoenvironmental scenario, located near an old lagoon-shelf outlet with a very strong marine influence and a mixture of possible polluted inputs by tidal inputs; b) the very low metal contents in comparison with core A; and c) a geochemical composition clearly linked to the grain size distribution of each sample, independently of the historical period.

Consequently, the greatest concentrations of metallic pollutants were found in clayey, inner lagoon sediments ($Fe_2O_3 > 4\%$; $Al_2O_3 > 11\%$; As > 10 ppm; Cu > 30 ppm; Pb > 30 ppm; Zn > 60 ppm), whereas sediment contamination by heavy metals declines significantly in sandy, aeolian or tsunami deposits (Fe_2O_3 < 1%; Al_2O_3 < 3%; As < 8; Cu < 10 ppm; Pb < 10 ppm; Zn < 12 ppm). This distribution is usual in coastal sediments, with a concentration of heavy metals in the clayey-silty fraction due to the increase in specific surface properties of this fraction (De Gregori et al., 1996; Bouezmarni and Wollast, 2005).

4.3. Measure of Metal Pollution

In the Huelva estuary, both absolute values and I_{geo} permit to evaluate the heavy metal pollution caused during the different historical mining periods. During the Phoenician period, this area was unpolluted, with geoaccumulation indices below or slighty above background levels. The Roman period caused a first heavy metal enrichment, with a low to moderate pollution (Cu: $I_{geo} > 2.5$; Pb: $I_{geo} \sim 2$). Nevertheless, the last period of increasing mining activities (1870-1950) caused a very high heavy metal enrichment, with I_{geo} up to 5 for Cu and Zn (sample A-8). In addition, these sediments are two (e.g. Pb) to nine twices (e.g. Zn) more polluted than those deposited during the Roman times, according to the absolute values of these elements. These data coincide with those observed by Ruiz et al. (1997) in other cores collected along the Huelva estuary.

These differences are not observed in a more external area of the Guadalquivir (e.g. core B). In this core, values of I_{geo} are lower than 1.3 for all samples and elements. Consequently, all samples are unpolluted or present a very low pollution. This decrease of heavy metal pollution has been observed in the marine areas of polluted estuaries (Fernández Caliani et al., 1997).

CONCLUSION

In the Huelva estuaries (SW Spain), the geochemical consequences of different historical phases of pollution depend on the palaeoenvironmental scenario. In the inner, more protected areas (e.g. core A; Huelva estuary), these effects are clearly separated, with unpolluted sediments during Phoenician times (> 2600 years BP), a very low pollution derived from Roman activities and very high heavy metal concentrations caused by the last mining period (1870-1930). In the Doñana spit, located near the Guadalquivir River mouth, sediments remain unpolluted during these intervals, owing probably to the tidal control and the increasing distance from the pollutant sources. In this case (core B), grain size is the main factor controlling the element concentrations of sediments.

ACKNOWLEDGMENTS

This work was funded by the DGYCIT Project CGL2010-15810, an Andalusia Excellence Project funded by FEDER (SEJ-4770) and three Research Groups of the Andalousia Board (RNM-238, RNM-293 and RNM-349).

REFERENCES

AMA (1993) (Agencia de Medio Ambiente). Plan de policia de aguas del litoral andaluz. Junta de Andalucia, *unpublished report*, 132 pp.

Blanco Freijero, A. and Rothenberg, B. (1981). Exploración arqueometalúrgica de Huelva. Labor, Barcelona.

Borrego, J., Morales, J. A. and Pendón, J. G. (1993). Holocene filling of an estuarine lagoon along the mesotidal coast of Huelva: the Piedras River mouth, southwestern Spain. *J. Coastal Res., 9,* 242-254.

Borrego, J., Ruiz, F., González-Regalado, M. L., Pendón, J. G. and Morales, J. A. (1999). The Holocene transgression into the estuarine central basin of the Odiel River mouth (Cádiz Gulf, SW Spain): lithological and faunal assemblages. *Quat. Sci. Rev. 18,* 769-788.

Bouezmarni, M. and Wollast, R. (2005). Geochemical composition of sediments in the Scheldt estuary with emphasis on trace metal. *Hydrobiologia, 540*, 155-168.

De Gregori, I., Pinochet, H., Arancibia, M. and Vidal, A. (1996). Grain size effect on trace metal distribution in sediments from two coastal areas of Chile. *Bull. Env. Contam. Toxicol., 57*, 163-170.

Fernández Caliani, J. C., Ruiz, F. and Galán, E. (1997). Clay mineral and heavy metal distribution in the lower estuary of Huelva and adjacent Atlantic shelf, SW Spain. *Sci. Total Env., 198*, 181-200.

Ferrero, M. D. (1988). Los conflictos de Febrero de 1888 en Riotinto. *Huelva en su Historia, 2*, 603–623.

Freitas, M. C., Andrade, C., Moreno, J. C., Munhá, J. M. and Cachao, M. (1999). The sedimentary record of recent (last 500 years) environmental changes in the Seixal Bay marsh, Tagus estuary, Portugal. *Geol. Mijnb., 77,* 283–293.

IEO (1992) (Instituto Español de Oceanografía). Variation espacio-temporal de parámetros físico-químicos y biologicos en la ria de Huelva y área de influencia, en el periodo 1987-1991. *Internal report no. 138*, 103 pp.

Köster, D., Lichter, J., Lea, P. D. and Nurse, A. (2007). Historical eutrophication in a river–estuary complex in mid-coast Maine. *Ecol. Appl., 17,* 765–778.

Pozo, M., Carretero, M. I., Ruiz, F., Rodríguez Vidal, J., Cáceres, L. M. and Abad., M. (2008). Caracterización mineralógica de facies sedimentarias de edad Pleistoceno superior-Holoceno en el Parque Nacional de Doñana (Huelva). Implicaciones paleoambientales. *Geotemas, 10,* 953-956.

Ruiz, F., González-Regalado, M. L., Borrego, J. and Morales, J. A. (1997). The response of ostracod assemblages to recent pollution and sedimentary processes in the Huelva Estuary, SW Spain. *Sci. Tot. Env., 207*, 91-103.

Ruiz, F., Rodríguez-Ramírez, A., Cáceres, L.M., Rodríguez Vidal, J., Carretero, M.I., Clemente, L., Muñoz, J.M., Yañez, C and Abad, M. (2004). Late Holocene evolution of the southwestern Doñana Nacional Park (Guadalquivir Estuary, SW Spain): a multivariate approach. *Palaeogeog., Palaeoclimatol., Palaeoecol., 204,* 47-64.

Ruiz, F., Rodríguez-Ramírez, A., Cáceres, L. M., Rodríguez Vidal, J., Carretero, M.I., Abad, M., Olías, M. and Pozo, M. (2005). Evidence of high-energy events in the geological record: Mid-Holocene evolution of the southwestern Doñana National park (SW Spain). *Palaeogeog., Palaeoclimatol., Palaeoecol., 229*, 212-229.

Ruiz, F., Abad, M., Galán, E., González, I., Aguilá, I., Olías, M., Gómez Ariza, J. L. and Cantano, M. (2006). The present environmental scenario of El Melah Lagoon (NE Tunisia) and its evolution to a future sabkha. *J. African Earth Sci., 44,* 289-302.

Ruiz, F., Borrego, J., López-González, N., Abad, M., González-Regalado, M.L., Carro, B. and Pendón, J. G. (2007). The geological record of a Mid-Holocene marine storm in southwestern Spain. *Geobios, 40,* 689-699.

Soares, A. M. M. and Martins, J. M. M. (2009). Radiocarbon dating of marine samples from Gulf of Cadiz: The reservoir effect. *Quat. Int., 221*, 9-12.

Zazo, C., Goy, J. L., Hillaire-Marcel, C., Dabrio, C. J., Belloumini, G., Improta, S., Lario, J., Bardají, T. and Silva, P. G. (1994). Holocene sequence of sea-level fluctuations in relation to climatic trends in the Atlantic-Mediterranean linkage coast. *J. Coast. Res., 10,* 933-945.

In: Heavy Metal Sediments
Editor: Lucía H. Sanz
ISBN: 978-1-61470-690-8
© 2012 Nova Science Publishers, Inc.

Chapter 6

HEAVY METAL CONTAMINATION CAUSED BY MINING ACTIVITIES IN THE MAR MENOR LAGOON (SE SPAIN)

Héctor M. Conesa-Alcaraz,[1]
Francisco J. Jiménez-Cárceles,[2*]
Antonio María-Cervantes,[1]
María Nazaret Gonzalez-Alcaraz,[1]
Consuelo Egea-Nicolás,[1]
and José Álvarez-Rogel[1]

[1]Departamento de Ciencia y Tecnología Agraria.
Escuela Técnica Superior de Ingeniería Agronómica. Universidad Politécnica de Cartagena. Paseo Alfonso XIII, Cartagena, Murcia, Spain
[2]BIOCYMA, Consultora de MedioAmbiente y Calidad.
Polígono Industrial La Serreta. Molina de Segura, Murcia, Spain

ABSTRACT

The *Mar Menor* lagoon is one of the most unusual ecosystems in the Mediterranean area. Many economic and industrial activities concur in the zone and the combined impacts of mining, agriculture and urban

---

* www.biocyma.com.

development in the surroundings of the lagoon during recent decades have affected the ecosystem. In this chapter we review the impacts of mining activities on the lagoon, and report data on contamination levels and the effects generated in the lagoon ecosystem. The discharge of wastewaters from mineral refining processes appears to be the most important source of heavy metal inputs to the lagoon. Efforts to decrease these inputs must be made at the headwaters of streams where hundred of hectares of tailings remain unprotected against erosion.

1. Introduction

Several authors have pointed out the role of wetland soils as sink for metals (Mays and Edwards 2001; Cheng et al., 2002), although under some environmental conditions, such soils may act as source of leachates with high metal concentrations (Weis et al., 2002; Peltier et al., 2003). The latter increases the bioavailability for fauna and flora and may represent a risk for metal accumulation in the food chain.

Metal solubility in inland soils is known to be controlled by pH. Neutral-basic conditions favour precipitation of metals as oxides and hydroxides. Thus, in oxic soils, metal dynamics are controlled by adsorption on organic matter, clay surfaces, carbonates, phosphate and oxide-hydroxide compounds (Van der Berg. 1999).

However, the factors that control metal dynamics in wetlands include several interrelated parameters such as pH, redox potential, salinity, organic matter, oxides, hydroxides, carbonate and microbial activity. Amongst these, redox potential (Eh) and pH have the greatest effect on metal transformations. Changes in Eh may directly affect the oxidised form of the metal (e.g. under oxidized conditions Fe^{2+} is transformed into Fe^{3+}). On the other hand, modifications in Eh may affect the stability of Fe and Mn oxide-hydroxides, thereby favouring the adsorption or desorption of metals (Kabata-Pendias, 2001). It is assumed that maximum metal solubility occurs when Eh values are low enough (<300 mV) for reduction of Fe and Mn oxides and hydroxides, thus releasing metals into the soil solution, but high enough (>100 mV) to prevent the formation of insoluble metal sulphides (Jacob and Otte, 2003; Figure 1). Greater accumulation of metals, due to the formation of metal sulphides in the deepest soil layers in zones with low Eh would therefore be expected.

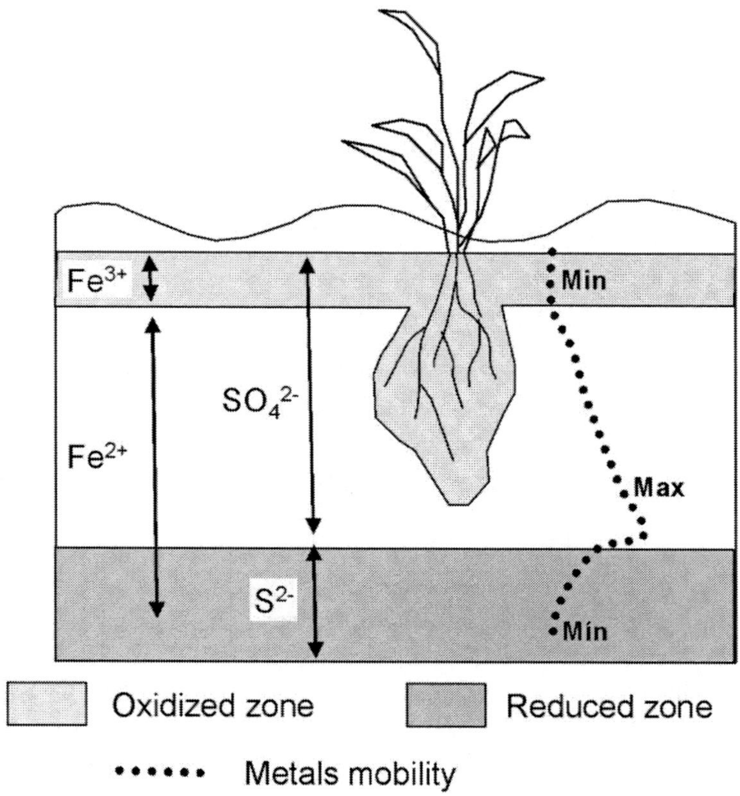

Modified from Jacob and Otte, 2003.

Figure 1. Diagram showing the relationship between vegetation, the oxidation state of soil and mobility of metals in hydric soils.

Plants can modify the soil in the rhizosphere zone and thus influence the dynamics of trace metals. A diagram of the soil-plant system in wetland soils is shown in Figure 1. The upper soil layer is oxidized due to the oxygen diffusion from the air, and the rhizosphere is oxidized due to the diffusion of oxygen from the plant roots. Metals are precipitated in these zones as oxides and hydroxides. In the lower anoxic soil layers, metals are also immobilized, in this case in the form of sulphides. However, metals display maximum mobility at positions with intermediate values of Eh, where oxides have dissolved but not yet formed sulphides.

The main form of copper in aerobic and alkaline conditions is $CuCO_3$, although Cu^{2+} and $CuOH^+$ and Cu(OH) complexes are also present. Copper sulphide (CuS) is formed under anaerobic conditions and in the presence of sulphur (Evanko and Dzombak, 1997).

Under most conditions, Pb^{2+} and lead hydroxide-complexes are the most stable forms (Smith et al., 1995). Lead carbonates are formed at pH= 6, and PbS is the main form in the presence of high concentrations of sulphur and reducing conditions.

At high pH, zinc may form carbonates and hydroxide complexes, which determine its solubility. Zinc precipitates under reducing conditions and, when present at high concentrations, may co-precipitate as iron and manganese oxyhydroxides (Smith et al., 1995).

The most common forms of cadmium in soil are as Cd^{2+}, cadmium-cyanide complexes, and $Cd(OH)_2$ (Smith et al., 1995). Hydroxides ($Cd(OH)_2$) and carbonates $(CdCO)_3$ are present at high pH. Under reducing conditions and in the presence of sulphur, cadmium sulphide is formed as a stable form. Cadmium also precipitates with phosphate, arsenate, chromate and other anions, although its solubility is strongly influenced by pH (Evanko and Dzombak, 1997).

Arsenic is a metalloid that under aerobic conditions appears mainly in the form of As(V), especially as arsenate (AsO_4^{3-}). Arsenate may act as a chelating agent and precipitate in the presence of other cations (Bodek et al., 1988). As (V) it may co-precipitate with iron oxyhydroxides under acid conditions. The mobility of these compounds increases as pH increases (Smith et al., 1995). Under reducing conditions As(III) is the main form, as arsenite (AsO_4^{3-}). Arsenite may precipitate with sulphur and displays a high affinity for sulphur compounds. Arsine (AsH_3) and As are present under strongly reducing conditions.

However, although the mobilisation of heavy metals in saltmarsh soils is strongly influenced by the transformation of oxides, hydroxides and sulphides, the thermodynamic properties of each metal may be more important than the concentration in the soil reactive fraction or even the redox status (Otero and Macías, 2002). Hence, detailed studies may be necessary in order to characterize the behaviour of each metal at a specific site.

2. An Overview of the Mar Menor and Surrounding Territories

The Mediterranean coast shelters a substantial human population and is a suitable zone for agriculture, industry and tourism. Some areas have of the Spanish Mediterranean coast have been mined during centuries, thus generating an additional impact on the regional ecosystems. Coastal lagoons in these areas frequently display physical and chemical disturbances and fluctuations, and are sometimes considered as waste sinks, which cause their accelerated environmental degradation (Pérez-Ruzafa et al., 1991, 2006).

The *Mar Menor* lagoon is one of the largest hypersaline coastal lagoons in the Mediterranean Sea. It is located in the southeast of Spain, in the eastern area of the Betic Ranges, and belongs to the geographical unit of *Campo de Cartagena*, a quaternary plain located in the southeast of the Region of Murcia (Latitude: 37°43'N; Longitude: 00°48'W) (Figure 2). Mining was the main economic activity in the area for centuries. Nowadays, the area is an important centre for agriculture, fisheries and tourism.

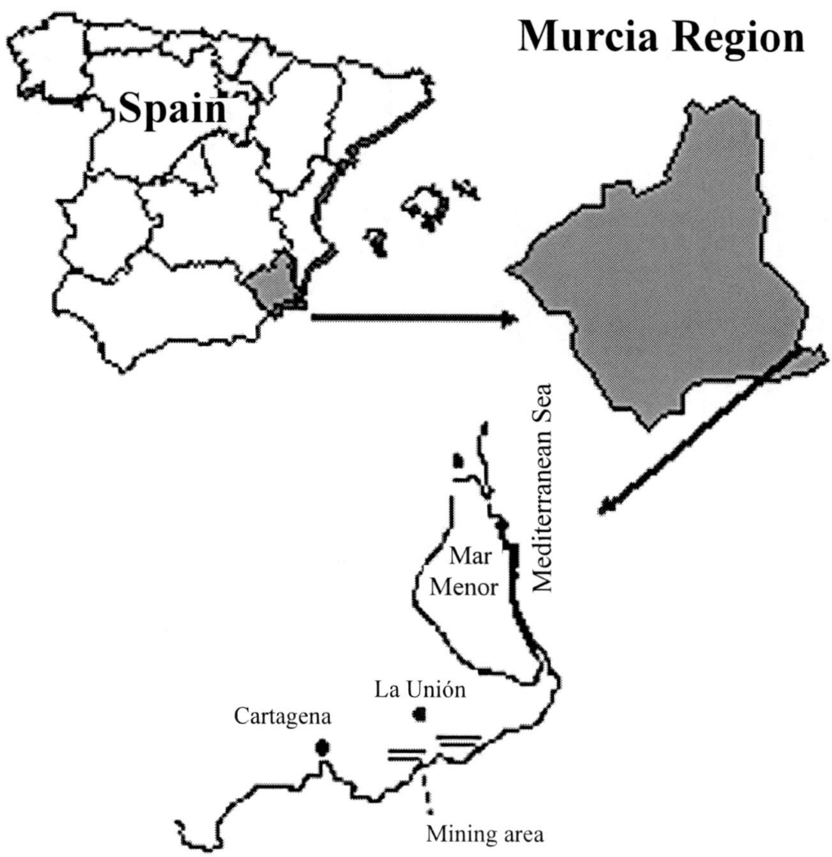

Figure 2. Map showing the location of the Mar Menor lagoon and the mining area.

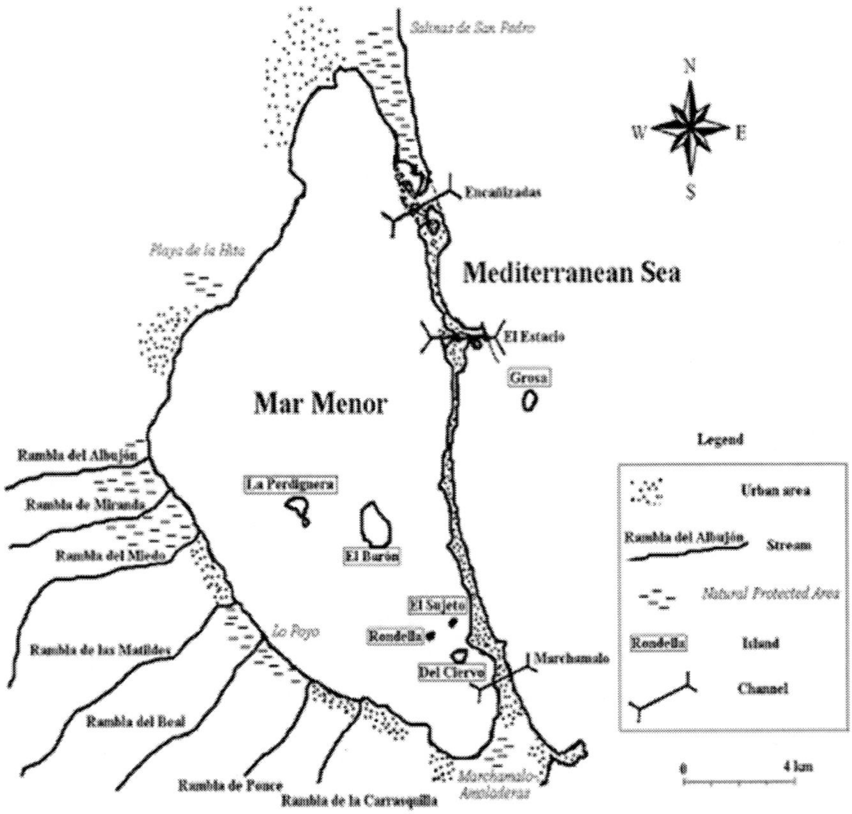

Figure 3. Map of the *Mar Menor* lagoon showing the main surface watercourses (*ramblas*), islands and channels.

The lagoon is triangle shaped, of size 13 x 15 x 21 km, mean depth of 2.5 m and maximum depth of of 6 m. It is separated from the Mediterranean Sea by a 20 km wide sand coastal barrier called *La Manga del Mar Menor.* There are some zones of communication between the lagoon and the Mediterranean Sea: a natural channel located in the North of the lagoon, and two artificial channels. There are five islands within the lagoon: *del Barón*, *Perdiguera*, *del Sujeto*, *Redondella* and *del Ciervo*.

The Mar Menor receives water run-off from the streams of the Sierra of Cartagena and La Unión and from the precoastal plain of *Campo de Cartagena* (Figures 2 and 3). These water streams or creeks, called *ramblas* are characteristic of the Mediterranean areas and carry water only occasionally, during the short, intense rainfalls typical of the Mediterranean climate.

The shoreline inside the lagoon is formed by a low lying sandy coast. Salt flats and saltmarshes occur at the mouths of the *ramblas* and in the inland depressions, and are surrounded by narrow dune systems. These wetlands originate from former secondary lagoons that were isolated by sediments carried to the basin, and undergo periodic phases of drying-watering due to variations in the groundwater level (ESA, 2003). These areas are the most important from an ecological point of view because of the presence of autochthonous flora and the fauna. The most important natural area in the lagoon, the *Salinas de San Pedro* Natural Park, is located at the northern end of the lagoon and covers an area of 822 ha (Figure 3). The smaller *Playa de la Hita* saltmarsh (70 ha), *Marina del Carmolí* saltmarsh (242 ha), *Lo Poyo* saltmarsh (186 ha) and *Marchamalo-Amoladeras* natural area (98 ha) are located in western and southern areas of the lagoon (ESA, 2003).

The lagoon and saltmarshes include 15000 ha protected by official laws and conventions; the area was designated as a Ramsar international site in 1994 and includes five Sites of Community Importance (SCI). It is considered a Special Protected Area of Mediterranean Interest (SPAMI), established by the Barcelona Convention in 2001, and a Site of Community Importance (SCI) to be integrated in the Nature 2000 Network (EU Habitats Directive). This zone is also a Specially Protected Area (SPA) in relation to nest building, migration and hibernation of aquatic birds, and is protected by European legislation (Birds Directive 79/409/CEE). The most important bird species include: avocet (*Recurvirostra avosetta*), black-winged stilt (*Himantopus himantopus*), snowy plover (*Charadrius alexandrinus*), and little tern (*Sterna albifrons*). Activities in the *Mar Menor* and its saltmarshes are officially regulated at local (Law 4/92 of the Regional Government of Murcia), national (Law 4/1989 of the Spanish Government) and international levels (Special Protected Area of Mediterranean Interest, SPAMI).

The climate in the zone is typically Mediterranean, semiarid, with warm summers and moderate winters and an annual average temperature of around 17°C. The mean annual rainfall is approximately 300 mm, mainly concentrated in October and April. The regional climatic conditions result in a natural flora adapted to drought and high temperatures. Most of the natural vegetation formations are degraded due to anthropogenic disturbance, and is only found on hills and in small areas, and are represented by small formations of pine trees (*Pinus halepensis* Miller) and groups of typically Mediterranean brushes such as esparto grass (*Stipa tenacísima* L.), palm (*Chamaerops humilis* L.), kermes oak (*Quercus coccifera* L.), mastic tree (*Pistacea lentiscus* L.) and aromatics such as rosemary (*Rosmarinus officinalis* L.) and thyme

(*Thymus hyemalis* Lange) (Martos-Miralles et al., 2001). Some plant species adapted to saline conditions grow in the coastal areas, including *Arthrocnumum macrostachyum* (Moric.) Moris and *Sarcocornia fruticosa* (L.) A.J. Scott, *Juncus maritimus* Lam., *Juncus acutus* L., *Scirpus holoschoenus* L., *Lygeum spartum* L., *Suaeda vera* Forsskal ex J.F. Gmel., *Sporobolus pungens* (Schreber) Kunth, *Elymus farctus* (Viv.) Runemark ex Melderis and *Pancratium maritimum* Lag. (Álvarez-Rogel, 1997; Álvarez-Rogel et al., 2001 and 2006).

The original soils in the zone are basic, as a result of the high concentration of $CaCO_3$. The most common type is Haplic Calcisols (WRB, 2006) or Haplocalcids (Soil Taxonomy, 1999) if the $CaCO_3$ is not cemented or Petric Calcisols (WRB, 2006) or Petrocalcids (Soil Taxonomy, 1999) if there is a petrocalcic horizon. Near the coast, Arenosols (WRB, 2006), or Psamments (Soil Taxonomy, 1999), are present in sandy zones and dunes, and Solonchaks (WRB, 2006) or Salids (Soil Taxonomy, 1999), are present in salt marshes.

The ecological equilibrium of the lagoon ecosystem is threatened by the massive urban growth in the surrounding areas, the intensive agricultural activity and the remains of the former mining activities in the nearby Cartagena - La Unión area. The objective of the present study was to review the environmental impacts of mining activities on the lagoon ecosystem.

3. Mining Activities in the Area: History

Mining activities in the Cartagena-La Union Mining District were for a long time the most important economic activity in this small area in the Murcia Region (southeast Spain). The presence of the industry led to the development of socio-economic networks based on metal mining and related processes: refining centers, smelters, etc. The Sierra Minera crosses the Mining District from west to east, parallel to the Mediterranean coast and belongs to the eastern part of the Betic Ranges, which were generated by alpine folding. The terrain is low-lying (<400 m), but with high slopes because of the proximity to the coast. The ore deposits in this zone contain iron, lead and zinc as the main metal components. Iron is present as oxides, hydroxides, sulphides, sulphates, carbonates, and silicates; lead and zinc are present as galena, sphalerite, carbonates, sulphates, and lead- or zinc- bearing (manganese, iron) oxides (Oen et al., 1975). The mining district covers an area of approximately 50 km^2, including five population nuclei with around 20,000 inhabitants.

Mining began during the Iberian period and continued with the Phoenicians, Carthaginians, etc., and reached a crucial point during the Roman domain (209 B.C.). Cartago-Nova (nowadays Cartagena) was one of the most important Roman cities in ancient Hispania. The Romans improved the extraction of iron, lead and silver. Mining activity during Roman times was so intensive that after the decline of the Roman Empire, the wastes generated were used for centuries as a source of minerals. The mining activity decreased after the decline of the Roman Empire because of depopulation of the area, among others reasons (Manteca and Berrocal, 1997), although a low level of activity was maintained in the following centuries.

The middle of the XIX century marked an inflexion point in the region because of some official systems that favoured mining activities (S.M.M.P.E., 1983). In addition, the development of metallurgy and new smelting techniques provoked a "boom" in mining activities in southeast Spain (Egea-Bruno, 2003). The mining was based on several small underground exploitations spread through galleries.

The mining activity acquired a new dimension in the middle of the XX century due to the concentration of mine properties at the multinational company Peñarroya-España S.A. This company introduced large new open cast mines (more efficient for the mining of the low metal ores) and the new refineries that applied differential flotation processes, which enabled larger quantities of material to be processed.

The total metal production increased until the early 1980s when the exhaustion of ore reserves, the low metal prices in the international market, the progressive elimination of the public protectionism to the mining sector, and the growing awareness of the environmental problems, resulted in a deep crisis in the local mining sector (Egea-Bruno, 2003). The mining activities in this region finally came to end in November 1991.

4. Environmental Impacts of Mining in the Lagoon

Mining activities are well known for their deleterious effects on the environment, particularly because of the presence of high concentrations of heavy metals such as copper, zinc, cadmium, lead and arsenic, which are widely known for their adverse effects on the environment and human health.

The negative effects of mining and related activities (refining, smelting) were noted during the 19th century when the wastes covering the riverbeds that flow into Mar Menor lagoon formed deposits of more than 3 metres thick (Vilar and Egea Bruno, 1990). Wastes resulting from refining processes were traditionally dumped into the streams that drained the local mountains ranges (locally called Sierra) and that mainly flow into the *Mar Menor* lagoon. The large quantities of wastes in the streams of the mountain caused the occasional flooding that occurs in the area to spread the wastes to croplands and shores.

In 1955 the Spanish Government prohibited the mining companies from dumping wastes on the ground to form mining tailings (Vilar et al., 1991). Nevertheless, with the recurrent torrential rains typical of the Mediterranean climate, the tailings that remain in the mountains continue to be eroded and washed into the lagoon (Marín-Guirao et al., 2005b).

The total area of the mining district is 50 km^2, and we can affirm that almost 20% of the surface is directly affected by the mining activity. This does not include the surroundings and other affected zones (continental platform, beaches, riverbeds). The current situation in the Sierra can be summarised, according to Martinez-Orozco et al. (1993), as follows:

- 600 ha covered by spoil piles (non-refined wastes from open cast mines)
- 220 ha affected by open cast mines.
- 160 ha covered by mine tailings

Most of these areas have not undergone any environmental restoration, and suffer erosion problems because the bare surfaces are exposed to erosive agents. The short intensive rainfall, typical of this zone may also relocate large volumes of material from these sites to the nearby lagoon. Chemical data on wastes located in the headwater of local streams are summarised in Table 1. Conesa et al. (2006) studied the concentrations of metal in tailings in the headwater of some *ramblas* (streams) in the area, and reported values of: 7670-11600 mg kg^{-1} Zn, 94-530 mg kg^{-1} Cu and 4900-7900 mg kg^{-1} Pb. The same authors concluded that approximately 14% of Zn was readily soluble in water in some of these substrates, which indicated the high risk of metal leaching in the run-off waters. Most of the *ramblas* in the Sierra of Cartagena-La Unión flow into the lagoon. Robles-Arenas et al. (2006) studied run-off waters from the Sierra, and reported mean concentrations of 77 mg L^{-1} Zn, 1.3 mg L^{-1} Cd, and 0.54 mg L^{-1} Pb. Because of the large quantities of waste in these *ramblas*, occasional flooding may spread the wastes to croplands and coastal environments. García-García (2004) analysed topsoil samples from the *Rambla del Beal* and *Rambla de Ponce* (Figure 3) flood plains and reported mean total concentrations of 4000 mg kg^{-1} Pb, 4600 mg kg^{-1} Zn, 4.5 mg kg^{-1} Cd, and 50 mg kg^{-1} Cu.

Some authors have studied the presence of metal-rich sediments from the mining activities in the lagoon (Table 2). Simonneau (1973) (cited in Martos-Miralles et al., 2001) estimated that there are 25 million tons of mining sediments in the lagoon. The highest concentrations of metals are reached in the southern area of the lagoon, which coincides with the mouths of the *ramblas* that drain the mining area. In these sites, the acid run-off water (pH 2-4) of these streams contacts with the alkaline water of the lagoon (pH > 8) and some of the metals in solution precipitate. A second distribution of metals takes place in the suspended particles and colloids, which are carried by the sea currents to the middle of the lagoon (García-García, 2004). Álvarez-Rogel et al. (2004) and Jiménez-Cárceles et al (2006) reported the presence of high concentrations of heavy metals (mainly Pb and Zn) in some coastal salt-marshes around the lagoon (Table 1). These authors concluded that there was a high risk of spreading the pollution to nearby agricultural and recreational areas.

The concentrations of metals that these authors reported widely surpass the threshold values proposed by some European legislation concerning soil remediation (MHSPE, 2000). The values are also much higher than those reported by Simón et al. (1999) -360 Pb mg kg^{-1}, 640 mg kg^{-1} Zn- found after the toxic spill in Aznalcollar (southern Spain) in 1998. Some intervention levels were proposed for the nearby region of Andalucía and after the toxic spill in 1998 (BOJA, 1999). This accident was considered as one of the major ecological disasters in Europe in recent decades. The maximum levels proposed in the legislation for natural areas (1200 mg kg^{-1} Zn, 10 mg kg^{-1} Cd, 500 mg kg^{-1} Pb) are widely surpassed at the mouths of the local *ramblas* that flow into the *Mar Menor* lagoon. Despite the high levels of metal pollution in this area, no studies regarding the bioavailability or metal uptake have been carried out in the nearby agricultural areas. Marin-Guirao et al. (2005a) and María-Cervantes et al. (2009) determined the levels of metals in the lagoon sediments and reported high concentrations of up to 7000 mg kg^{-1} Zn and Pb and 8 mg kg^{-1} Cd.

Table 1. Metal concentrations reported for some sites in the Cartagena-La Unión Mining District at the headwaters of local ramblas and comparison with reference levels. All data are in mg kg^{-1}; n.a.: no data available

Headwater site	Water stream	Reference	Lead	Copper	Zinc	Arsenic	Cadmium
Mine tailing "Belleza"	Rambla del Miedo and Rambla de las Matildes	Conesa et al. (2008)	7000	380	5400	1900	8.8
Mine tailing "El Lirio"	Rambla de la Carrasquilla	Conesa et al. (2007a)	4800	170	Up to 15000	n.a.	n.a.
Mine mailing "Brunita"	Rambla del Miedo	Conesa et al. (2007b)	2200	1100	2000	n.a.	n.a.
Some reference legislation	Maximum metal levels allowed in soils following the Aznalcollar accident (Spain)	B.O.J.A. (1999)	500	500	1200	100	10
	Mean concentration reported for soils affected by the Aznalcollar toxic spill	Simón et al. (1999)	361	152	642	121	2.3

The presence of metals in the sediments of the lagoon has affected the organisms that live in this area. Auernheimer et al. (1984, 1996) observed high concentrations of metals in bivalve shells from the lagoon (up to 300 mg kg^{-1} Zn and 218 mg kg^{-1} Pb). Sanchiz et al. (2000) studied the sediments and macrophytes along the Spanish Mediterranean coast, including samples from the *Mar Menor*. The levels of Pb, Zn and Cd found in the sediments from *Mar Menor* lagoon (Table 2) were higher than those obtained at different points of the Spanish coast.

The metal uptake in macrophytes like *Cymodocea nodosa*, *Caulerpa prolifera*, and *Ruppia cirrhosa* were also studied (Table 2). The highest concentrations of Pb and Zn were found in *C. nodosa*, and the authors of the study suggested that this species may be useful for metal biomonitoring in the area. However, this study did not suggest any pathways that would enable conclusions to be reached about metal bioavailability.

Table 2. Heavy metal concentrations reported by some authors for sediments and water from the Mar menor lagoon and the surrounding *ramblas*. Soil data are expressed in mg kg^{-1} and water data in mg L^{-1}

	Site	Metal				Reference
		Cd	Pb	Cu	Zn	
Sediments in lagoon		0.31-0.60	30-90	n.a.	130	Sanchiz et al. (2000)
Sediments in lagoon		Up to 7	up to 10000	n.a.	up to 7000	Marin Guirao et al. (2005c)
Sediment Stream mouths	Rambla de la Carrasquilla	24	3690	39	3500	García-García (2004)
	Rambla de Ponce	25	1960	30	1990	
	Rambla del Beal	15	950	35	650	
	Rambla del Miedo	16	2250	42	2850	
Coastal saltmarsh soils	Lo Poyo saltmarsh	n.a.	8000	110	6940	Alvarez-Rogel et al. (2004)
	Marina del Carmolí saltmarsh	up to 100	up to 1500	up to 450	up to 60000	Jiménez-Cárceles et al (2006)
Free water	Rambla de la Carrasquilla	0.03	0.04	n.a.	0.05	García-García (2004)
	Rambla del Beal	0.02	0.07	n.a.	0.41	

n.a.: no data available.

Marin-Guirao et al. (2005b and 2005c) developed a more specific sampling method, and centered their study on the southern area of the lagoon (the most affected by metal pollution). These authors studied the metals in sediments and metal uptake in *C.nodosa* and reported higher values than Sanchiz et al. (2000) (Table 3). After studying the benthic communities from some sampling stations located at the mouth of the *ramblas,* and relating them with the physicochemical properties of the sediments, they found the highest correlations for metal concentrations. They concluded that the physicochemical properties have affected the growing of benthic communities in this part of the lagoon (Marin-Guirao et al., 2005b and 2005c). In the latter study (Marin-Guirao et al., 2005c), the authors also analysed the biofilm, defined as algae, bacteria and associated fine detrital material, which adheres to substrates in water bodies (Ruttner, 1968) and is a food source for invertebrates that feed on mineral and organic surfaces (Merrit and Cummins, 1978). This study determined that metals present in the biofilm may be used as an indicator of the long-term evolution of the contaminants in the sediments and reflect the potential bioavailability of metals for the biota. The values of metal concentration in biofilms of *C.nodosa* (Table 3) reflects the high bioavailability of the metals to the organisms that live in the lagoon, particularly for Zn (3900 mg kg^{-1}) and Pb (1500 mg kg^{-1}).

Data on metal accumulation in mollusks organisms from the lagoon are shown in Table 4. The data show high metal concentrations, particularly for Pb (up to 220 mg kg^{-1}) and Zn (up to 3130 mg kg^{-1}). These data may reflect the existence of a contamination in the food chain in relation to metals and constitute an important concern with regards to both animal and human health. It therefore appears necessary to control the fauna in order to prevent possible transfer of metals to the human diet, since commercial and leisure fishing are very popular in the area. More studies, particularly in relation to speciation and accumulation of metals in fish and other animals consumed by humans, are necessary in order to evaluate the associated health risks.

Table 3. Metal concentrations in macrophytes from the *Mar Menor* lagoon. Data are estimations from figures. Values are mg kg^{-1}

Species	Organ	Heavy metal			Reference
		Cd	Pb	Zn	
Caulerpa prolifera	Rhizoids+stolons	0.2	38	30	Sanchiz et al. (2000)
	Fronds	0.15	38	32	
Ruppia cirrhosa	Roots	0.5	38	56	
	Rhizomes+shoots	0.7	14	40	
	Leaves	0.4	30	70	
Cymodocea nodosa	Roots	0.18	38	30	
	Rhizomes	0.28	10	20	
	Leaves	0.2	126	78	
Cymodocea nodosa	Roots	0.5	850	500	Marin-Guirao et al. (2005b and 2005c)
	Rhizomes	0.15	250	200	
	Leaves	0.2	300	350	
	Biofilm	0.6	1500	3900	

Table 4. Heavy metal concentrations in mollusk species from the Mar Menor lagoon. Data are expressed in mg kg^{-1} dry weight. n.a.: no data available

Species	Heavy metal					Reference
	Fe	Cu	Cd	Pb	Zn	
Cerastoderma edule / glaucum	21.8	3.3	1.06	58	40	Simmoneau (1973)*
Venerupis aurea	n.a.	n.a.	1.5	172	40	
Cerastoderma edule glaucum	32.5	0.7	0.08	22.1	12.7	De Leon et al. (1982)
Venerupis aurea	50.8	2.4	0.14	75.6	10.4	
Cerastoderma edule / glaucum	up to 4000	n.a.	n.a.	up to 220	up to 300	Auernheimer et al. (1987 and 1996)
Venerupis aurea	up to 950	n.a.	n.a.	up to 75	up to 95	
Hexaples trunculus	n.a.	up to 47.2	up to 8.4	up to 222	up to 3130	María – Cervantes et al. (2009)

* Data cited in CEOTMA (1985).

CONCLUSION

The Mar Menor lagoon has received several tonnes of contaminated material throughout the last two thousand years of mining activities from the nearby area of Cartagena-La Unión.

Removal of the contaminated sediments in the lagoon appears to be difficult because of the huge quantities involved and the possible environmental impacts that the removal of sediments may have on the lagoon ecosystem. Nevertheless, the entrance of mining waste materials into the lagoon from the surrounding *ramblas* may be decreased by stabilization of the tailings at the headwaters and cleaning the contaminated riverbeds.

ACKNOWLEDGMENTS

We are grateful to the Spanish Ministry of Science (projects REN2001-2142/GLO , CGL2004-05807 and CGL2007-64915) and "Fundación Séneca" of the Comunidad Autónoma de la Región de Murcia for financial support.

Antonio María-Cervantes and Maria Nazaret González received of grants from the Spanish Ministry of Science. We thank Dr. Ripolles for his valuable comments and suggestions.

REFERENCES

Álvarez-Rogel, J. 1997. Relaciones suelo-vegetación en saladares del sureste de España. Ph. D. Thesis. Publicaciones CD-Rom Universidad de Murcia. Murcia, Spain.

Álvarez-Rogel, J., Martínez-Sánchez, J.J., Carrasco Blázquez, L., Marín Semitiel, C.M. 2006. A conceptual model of a salt marsh plant distribution in coastal dunes of Southeastern Spain. *Wetlands*, 26, 703-717.

Álvarez-Rogel, J., Ortiz, R., Vela de Oro, N., Alcaraz, F. 2001. The application of the FAO and US Soil Taxonomy systems to saline soils in relation to halophytic vegetation in SE Spain. *Catena*, 45, 73-84.

Álvarez-Rogel, J., Ramos-Aparicio, M.J., Delgado-Iniesta, M.J., Arnaldos-Lozano, R. 2004. Metals in soils and above-ground biomass of plants from a salt marsh polluted by mine wastes in the coast of the Mar Menor Lagoon, SE Spain. *Fresenius Environmental Bulletin*, 13, 274-278.

Auernheimer, C., Chinchon, S., Pina, J.A. 1996. Lead pollution in bivalve shells. Mar Menor, Spain. *Arch. Sc. Geneve*, 49, 87-98.

Auernheimer, C., Llavador, F., Pina, J.A. 1984. Chemical minority elements in bivalve shells. A natural model (Mar Menor, Spain). *Arch. Sc. Geneve* 37, 317-331.

Bodek, I., Lyman, W.J., Reehl, W.F., Rosenblatt, D.H. 1988. Environmental Inorganic Chemistry: Properties, Processes and Estimation Methods, Pergamon Press, Elmsford, NY.

BOJA (Boletín Oficial de la Junta de Andalucía). 1999. Orden de 18 de diciembre de 1998, por la que se fijan las concentraciones límites en los suelos afectados por el accidente minero de Aznalcóllar. (BOJA nº 5, de 12.01.99).

CEOTMA (Centro de Estudios de Ordenación del Territorio y Medio Ambiente). 1985. Ordenación territorial de la zona del Mar Menor y su entorno. Ministerio de Obras Públicas y Urbanismo. 130 pp.

Cheng, S., Grosse, W., Karrenbrock, F., Thoennessen, M. 2002. Efficiency of constructed wetlands in decontamination of water polluted by heavy metals. *Ecological Engineering,* 18, 317-325.

Conesa, H.M., Faz, Á., Arnaldos, R.2007a. Initial studies for the phyto-stabilization of a mine tailing from the Cartagena- La Union Mining District (SE Spain). *Chemosphere*, 66:38-44.

Conesa H.M, Faz, Á., García, G., Arnaldos, R. 2007b. Heavy metal contamination in the semiarid area of Cartagena-La Unión (SE Spain) and its implications for revegetation. *Fresenius Environmental Bulletin*, 16: 1076-1081.

Conesa, H.M., Faz, Á., Arnaldos, R. 2006. Heavy metal accumulation and tolerance in plants from mine tailings of the semiarid Cartagena–La Unión mining district (SE Spain). *Science of the Total Environment*, 366, 1-11.

Conesa, H.M., Robinson, B.H., Schulin, R., Nowack, B. 2008. Metal extractability in acidic and neutral mine tailings from the Cartagena-La Unión Mining District (SE Spain). *Applied Geochemistry,* 23, 1232-1240.

De León, A.R., Guerrero, J., Farazo, F. 1982. Evolution of the pollution of the coastal lagoon of Mar Menor. VI Journées Étud. Pollutions, Cannes, C.I.E.S.M. 355-358.

Egea-Bruno, P.M. 2003. La minería contemporánea en la Sierra Minera de Cartagena-La Unión. In: I. Rábano, I. Manteca and C. García, (Editors), Patrimonio geológico y minero y desarrollo regional, Instituto Geológico y Minero de España, Madrid, Spain, 31-42.

ESA (European Space Agency). 2003. Statement of Work Wetland. Annex A. Available on-line at: http://dup.esrin.esa.it/files/document/193-204-228-163_20031114114423.pdf (accessed 31 August 2006).

Evanko, C.R., Dzombak, D.A. 1997. Remediation of Metals-Contaminated Soils and Groundwater. Technology Evaluation Report. TE-97-01. Ground-Water Remediation Technologies Analysis Center, Pittsburgh, PA, US.

García-García, C. 2004. Impacto y riesgo medioambiental en los residuos minerometalúrgicos de la Sierra de Cartagena-La Unión. Ph. D. Thesis. Universidad Politécnica de Cartagena, Cartagena, Spain.

Jacob, D.L., Otte, M.L. 2003. Conflicting processes in the wetland plant rhizosphere: metal retention or mobilization? *Water, Air and Soil Pollution*, 3, 91-104.

Jiménez-Cárceles, F.J., Egea. C., Rodríguez-Caparrós, A.B., Barbosa' O.A., Delgado' M.J., Ortiz, R., Álvarez-Rogel, J. 2006. Contents of nitrogen, ammonium, phosphorus, pesticides and heavy metals, in a salt marsh in the coast of the Mar Menor lagoon (SE Spain). *Fresenius Environmental Bulletin*, 15, 370-378.

Kabata-Pendias, A. 2001. Trace elements in soils and plants. CRC Press, Florida, EEUU. 413 pp.

Manteca, J.I., Berrocal, M.C. 1997. Sobre la creación de un parque geominero y arqueoindustrial en La Unión (Murcia) como medida de protección y puesta en valor del patrimonio histórico-minero. II congreso empresarial C.O.E.C. Desarrollo Turístico-Económico de la comarca de Cartagena. Cartagena, 20-21 Febrero de 1997.

María Cervantes, A., Jiménez Cárceles, F.J., Álvarez Rogel, J. 2009. As, Cd, Cu, Mn, Pb and Zn contents in sediments and mollusks *Hexaples trunculus* and *Tapes decussatus* from coastal zones of a Mediterranean lagoon (Mar Menor, SE Spain) Affected by mining wastes. *Water, Air and Soil Pollution*. DOI 10.1007/s11270-008-9913-7.

Marín-Guirao, L., Cesar, A., Marín, A., Lloret, J., Vita, R. 2005a. Establishing the ecological quality status of soft-bottom mining-impacted coastal water bodies in the scope of the Water Framework Directive. *Marine Pollution Bulletin*, 50, 374-387.

Marín-Guirao, L., Cesar, A., Marín, A., Vita, R. 2005b. Assessment of sediment metal contamination in the Mar Menor costal lagoon (SE Spain): Metal distribution, toxicity, bioaccumulation and benthic community structure. *Ciencias Marinas*, 31, 413-428.

Marín-Guirao, L., Marín-Atucha, A., Lloret-Barba, J., Martínez-López, E., García-Fernández, A.J. 2005c. Effects of mining wastes on a seagrass ecosystem: metal accumulation and bioavailability, seagrass dynamics and associated community structure. *Marine Environmental Research*, 60, 317-337.

Martínez-Orozco, J.M., Valero-Huete, F., González-Alonso, S. 1993. Environmental problems and proposals to reclaim the areas affected by mining exploitations in the Cartagena mountains (southeast Spain). *Landscape Urban Plan*. 23, 195-207.

Martos-Miralles, P., Sansano Sánchez, A., Baños Páez, P., Navarro Cano, J.A., Méndez Pérez, T. 2001. Medio Ambiente y Empleo en la Sierra Minera de Cartagena-La Unión. Fundación Sierra Minera (eds). La Unión (Murcia), 256 pp.

Mays, P.A., Edwards, G.S. 2001. Comparison of heavy metal accumulation in a natural wetland and constructed wetlands receiving acid mine drainage. *Ecological Engineering*, 16, 487-500.

Merrit, R.W., Cummins, K.W. 1978. An introduction to the aquatic insects of North America. Kendall/Hunt Publishing Co., Dubuque, IA, 441 pp.

M.H.S.P.E. (Ministry of Housing, Spatial Planning and Environment). 2000. Netherlands. Circular on target values and intervention values for soil remediation. *Ministry of Housing, Spatial Planning and Environment*, 4-2-2000.

Oen, I.S., Fernández, J.C., Manteca, J.I. 1975. The lead-zinc and associated ores of La Unión, Sierra de Cartagena, Spain. *Econ. Geol.* 70, 1259-1278.

Otero, X.L., Macías, F. 2002. Variation in depth and season in metal sulfides in salt marsh soils. *Biogeochemistry*, 61, 247-268.

Peltier, E.F., Webb, S.M., Gaillard, J-F. 2003. Zinc and lead sequestration in an impacted wetland system. *Advances in Environmental Research* 8, 103-112.

Pérez-Ruzafa, A., García-Charton, J.A., Barcala, E., Marcos, C. 2006. Changes in benthic fish assemblages as a consequence of coastal works in a coastal lagoon: the Mar Menor (Spain, Western Mediterranean). *Marine Pollution Bulletin*, 53, 107-120.

Pérez-Ruzafa, A., Marcos-Diego, C., Ros, D. 1991. Environmental and biological changes related to recent human activities in the Mar Menor (SE of Spain). *Marine Pollution Bulletin*, 23, 747-751.

Robles-Arenas, V.M., Rodríguez, R., García, C., Manteca, J.I., Candela, L. 2006. Sulphide-mining impacts in the physical environment: Sierra de Cartagena–La Unión (SE Spain) case study. *Environmental Geology*, 51, 47-64.

Ruttner, F. 1968. fundamentals of limnology. University of Toronto. Presss, Canada, 295 pp.

S.M.M.P.E. (Sociedad Minero Metalúrgica Peñarroya España). 1983. Libro del centenario 1881-1981. Sociedad Minero Metalúrgica Peñarroya España, Madrid, Spain, 757 pp.

Sanchiz, C., García-Carrascosa, A.M., Pastor, A. 2000. Heavy metal contens in sofá-bottom marine macrophytes and sediments along the mediterranean coast of Spain. *Marine Ecology*, 21, 1-16.

Simón, M., Ortiz, I., García, I., Fernández, E., Fernández, J., Dorronsoro, C., Aguilar, J. 1999. Pollution of soils by the toxic spill of a pyrite mine (Aznalcollar, Spain). *Science of the Total Environment*, 242, 105-115.

Simonneau, J. 1973. Mar Menor: évolution sédimentologique et géochimique récente du remplissage. Ph. D. Thesis. Université de Toulouse. 172 pp.

Smith, L.A., Means, J.L., Chen, A., Alleman, B., Chapman, C.C., Tixier, J.S., Jr., Brauning, S.E., Gavaskar, A.R., and Royer, M.D. 1995. Remedial Options for Metals-Contaminated Sites, Lewis Publishers, Boca Raton, FL.

Soil Taxonomy. 1999. A Basic System of Soil Classification for Making and Interpreting Soil Surveys. Second Edition. US Dep. Agricult. Nat. Res. Cons. Serv. Agriculture Handbook 436, Washington, DC.

Van den Berg, G.A., Gustav, J.P., Van der Heijdt, L.M. and Zwosman, J.J.G. 1999. Mobilisation of heavy metals in contaminated sediments in the River Meuse, The Netherlands. *Water, Air and Soil Pollution*, 116, 567-586.

Vilar, J.B., Egea, P.M. 1990. La minería murciana contemporánea (1840-1930). Universidad de Murcia, Academia Alfonso X El Sabio, Excmo. Ayuntamiento de Cartagena, CajaMurcia, Spain, 389 pp.

Vilar, J.B., Egea, P.M., Fernández, J.C. 1991. La minería murciana contemporánea (1930-1985). Instituto Tecnológico Geominero de España, Madrid, España, 1991.

Weis, P., Windham, L., Burke, D. J., Weis, J. S. 2002. Release into the environment of metals by two vascular salt marsh plants. *Marine Environmental Research*, 54, 325-329.

WRB. 2006. World reference base for soil resources *2006.* World Soil Resources Reports No. 103. FAO, Rome.

In: Heavy Metal Sediments
Editor: Lucía H. Sanz

ISBN: 978-1-61470-690-8
© 2012 Nova Science Publishers, Inc.

Chapter 7

HEAVY METAL CONTAMINATION IN SELECTED URBAN COASTAL REGIONS IN US AND CHINA

Huan Feng,[1*] Weiguo Zhang,[2] Luoping Zhang,[3] Xu-Chen Wang,[4,5] Lizhong Yu,[2] and Danlin Yu[1]

[1]Department of Earth and Environmental Studies,
Montclair State University, New Jersey, US
[2]State Key Laboratory of Estuarine and Coastal Research,
East China Normal University, Shanghai, PRC
[3]State Key Laboratory of Marine Environmental Science,
Environmental Science Research Center, Xiamen University,
Xiamen, Fujian, PRC
[4]Department of Marine Ecology and Environmental Studies,
Institute of Oceanography, Chinese Academy of Sciences, Qingdao, PRC
[5]Department of Environmental,
Earth and Ocean Sciences, University of Massachusetts at Boston,
Boston, MA, US

* Corresponding author. Tel.: 973-655-7549; fax: 973-655-4072. E-mail: fengh@montclair.edu (H. Feng).

Abstract

With urbanization and economic development in coastal area, metal pollution in coastal environment has been a problem. Estuaries and coastal intertidal zone are important habitats for aquatic and marine life. In the meantime, sediment is a repository of contaminants and records the pollution history. Sediment quality reflects the long-term health status of an estuarine or coastal system and can be evaluated by sediment quality guidelines/criteria, metal enrichment factors and other indicators. In this study, we summarize the results mainly from our previous studies in metal pollution in sediments in urban estuarine and coastal systems in US and China, including New York Harbor and Yangtze River estuary in metropolitan areas and Xiamen Bay and Jiaozhou Bay. The information from this study expands our knowledge in understanding metal pollution in urban coastal systems and assessing environmental quality impacted by industrialization and economic development.

1. Introduction

Metals are naturally present in our environment and trace amount of metals is essential micronutrients for growth of many marine organisms. However, excessive amount of heavy metals can be toxic to marine life in the coastal environment (Underwood 1971; Langston 1990). With the urbanization and economic development, metal pollution in water and sediments in coastal areas has become a major environmental concern because the coastal area has a high population density and the estuaries and intertidal zone are important habitats for aquatic and marine life. Estuarine and coastal environment are considered as a complex system, where physical, chemical, and biological processes play major roles in behavior, transport and fate of metal contaminants. In coastal pollution study, anthropogenic impact is more apparent in estuarine and coastal area than that in open ocean because contaminants including toxic heavy metals are carried to the ocean by rivers via estuarine systems. Many heavy metals are transported predominantly in association with particulate matter that eventually settle down to the sediments, and consequently, high concentration of heavy metals are often detected in sediments in many industrialized harbors and coastal regions around the world because the major sources of anthropogenic metals to a coastal area are due to industrialization and urban development with population growth in the area (Forstner and Wittman 1983; Bothner et al.

1998; Feng et al. 2004; Santos et al. 2005). Therefore, estuarine and coastal sediments are repository for these pollutants that enter the sea and useful for environmental studies because it can provide time-integrated records (Goldberg et al. 1979; Trefry and Shokes 1981; Windom et al. 1989; Bopp and Simpson 1989; Valette-Silver 1993). This article summarizes our previous metal contamination studies in several coastal areas in the United States and China. It reflects the environmental concerns due to heavy metal contamination in these selected urban coastal regions in US and China with the urbanization and economic development.

2. Study Areas

2.1. Lower Hudson River - New York Harbor, USA

The Hudson River estuary is classified as a partially mixed estuary and the water sources to the New York Harbor (km point -11 to 18, as defined by Olsen et al. 1984) are mainly Hudson River water, New York Bight water and New Jersey/New York waste water (Abood 1978; Klinkhammer and Bender 1981; Cooper et al. 1988) (Figure 1). With the early industrial development in this urban coastal area, pollution becomes a serious environmental problem. Sediments in the New York Harbor are found to be significantly contaminated by heavy metals, PCBs and other contaminants (Williams et al. 1978; Olsen et al. 1984; Bopp and Simpson 1989; Hunt et al. 1994; Gibbs 1994; Chillrud 1996; Feng et al. 1998a).

Sedimentation rates in the Hudson River estuary were estimated by Oslen et al. (1978) using reactor- and bomb-produced radionuclides ($^{239,240}Pu$, ^{137}Cs, ^{134}Cs and ^{60}Co) as time-stratigraphic markers for the sediment record.

Relating these radionuclide profiles in the Hudson sediments to the known input histories of these radionuclides to the estuary, they found that fine-grained sediment accumulation rates are between 5 and 20 cm y^{-1} in most depositional areas of the New York Harbor. They also found that the fine-grained particle accumulation rates in some areas adjacent to New York City are generally >3 cm y^{-1} and often as high as 10 to 20 cm y^{-1}, assuming that bioturbation in the Hudson River estuary is relatively unimportant (Olsen et al. 1981). Feng et al. (1998b) estimated that sedimentation rates in the western margin area at kmp ~8 varied from 6 to 26 cm y^{-1} based on ^{7}Be profiles.

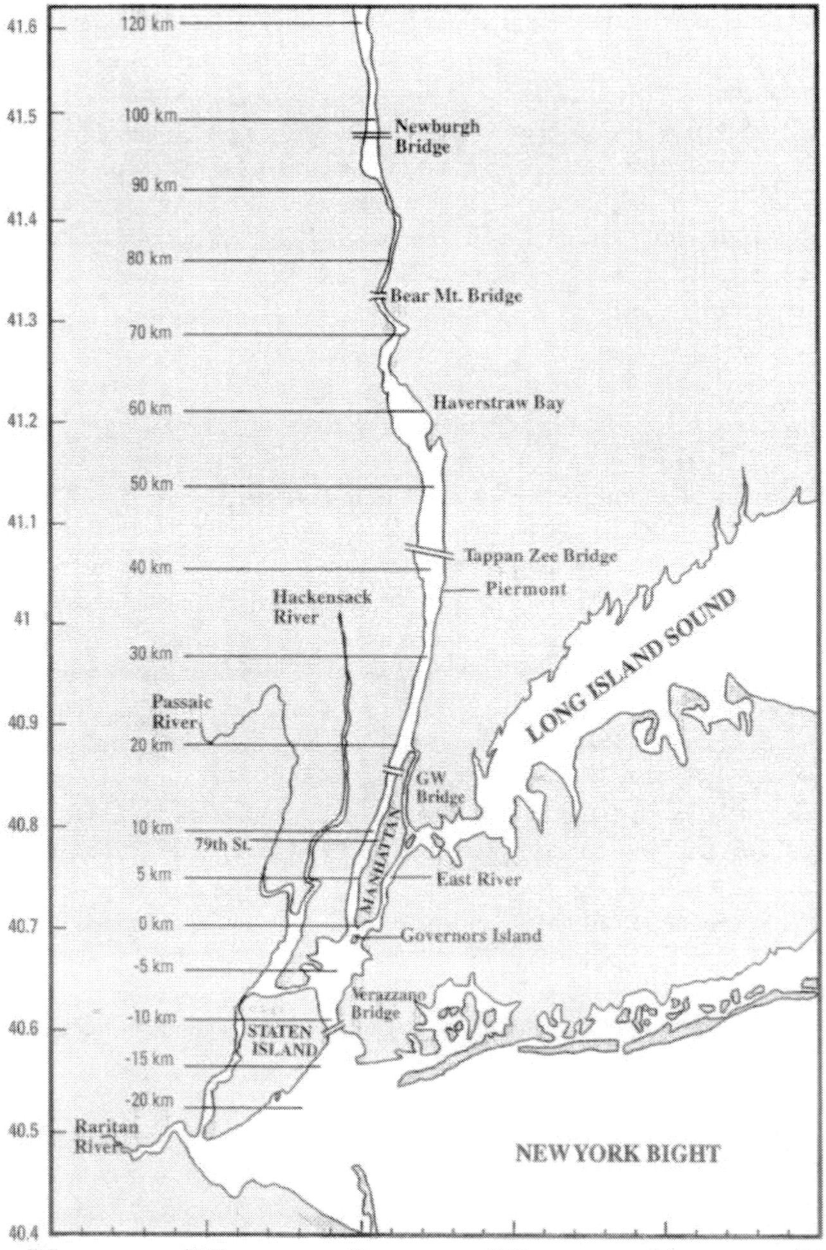

From Feng et al. 1998a.

Figure 1. Map showing the lower Hudson River estuary, USA.

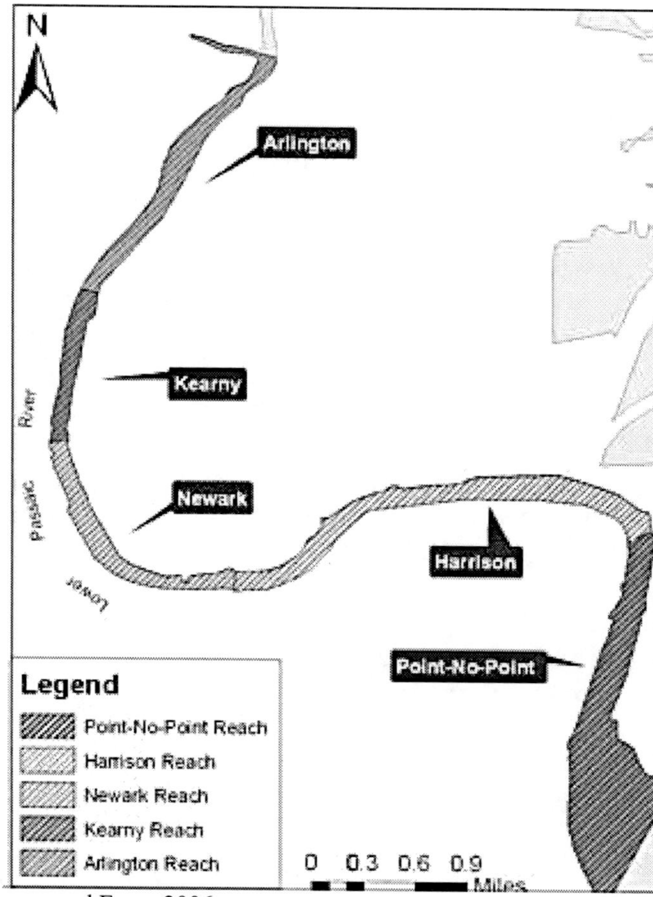

From Onwueme and Feng, 2006.

Figure 2. Map showing the lower Passaic River, New Jersey, USA.

2.2. Lower Passaic River, New Jersey, USA

The Passaic River is about 14 miles west to New York City, located in the New Jersey-New York metropolitan area. The Passaic River system consists of major and minor tributaries, and drains much of eight New Jersey counties as well portions of two New York counties and ultimately reaches Newark Bay at its confluence with the Hackensack River (Figures 1 and 2). The drainage basin of northern New Jersey's Passaic River system covers approximately 2200 km^2, about 11% of the state's surface area (NJDEP 1987). The basin contains some of the most densely populated land in the nation and is afflicted with numerous environmental problems, including major industrial pollution and suburban sprawl. This river has been heavily polluted by dioxins, PAHs, PCBs and heavy metals due to industrial activities and urbanization (Huntley et al. 1993, 1995, 1997; Iannuzzi et al., 2002; Onwueme and Feng 2006).

The Passaic River environmental status has been drawing much regional and national attention due to its urban environmental setting. The lower 10 km reach of the Passaic River is one of EPA Superfund Sites (Figure 2). In the lower Passaic River, fine-grained contaminated sediments have received considerable attention because these contaminants (metals, PCBs, PAHs, dioxins, etc.) have a potential to release into the aquatic system and air through diffusion, resuspension and/or evaporation, causing human health hazardous affect (Huntley et al. 1997; Walker et al. 1999; Wenning et al. 1994).

Since early 1980s, federal and New Jersey state agencies and private sectors have funded a series of remedial investigation of the lower 10 km of the Passaic River to characterize the horizontal and vertical distributions and concentrations of dioxins, PCBs, and metals in its sediment (Crawford et al. 1995; Finley et al. 1990; Hutley et al. 1993, 1995, 1997; Tierra Solutions, Inc. 2003; Wenning et al. 1993a, 1993b; Wolfskill and McNutt 2000). Numerous data were produced from these studies.

2.3. Yangtze River Estuary, China

The Yangtze River is the third largest river in the world and delivers about 4.7×10^8 tons of sediments to the sea annually (Chen 1998) (Figure 3). Therefore, intertidal zone in Yangtze River Estuary has been extensively well developed (Chen 1998). Previous studies show that heavy metal pollution has been found in some areas of the intertidal zone (Zhang et al. 2001; Feng et al. 2004). Shanghai, which is located along the Yangtze Estuary and one of the largest cities with most active industrial activities in China, has a direct impact on the coastal environment. It was reported that 5 million tons of industrial and domestic sewage were discharged into the Yangtze River estuary daily (Dai and Gu 1990). In addition to the local waste discharge, many areas in the intertidal zone have been used for harbors, iron and steel mills, and other heavy industries. Due to urbanization and economic development in past decades, metal contamination in the Yangtze River Estuary has received much attention (Xu et al. 1997; Chen et al. 2001; Feng et al. 2004; Zhang 1999; Zhang et al. 2001, 2007a).

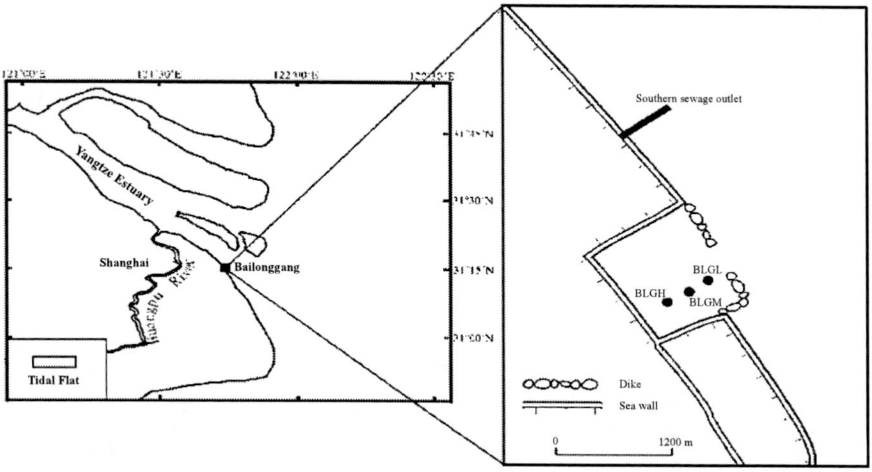

From Feng et al. 2004.

Figure 3. Map showing the study area in Yangtze River estuary intertidal zone, Shanghai, China.

2.4. Xiamen Bay, China

Western Xiamen Bay (24°29'N, 118°04'E) is a semi-enclosed bay in southeast China (Figure 4). The water depth ranges from 6-25 m with a deep-water coastline of about 30 km. Thus, it has an excellent natural condition for navigation and shipping activities. Although the economic development in Xiamen Economic Special Zone speeds up the urbanization and industrialization of the region in past 25 years, it has resulted in great environmental stress to western Xiamen Bay and its adjacent areas and brought environmental problems to the region.

2.5. Jiaozhou Bay, China

Jiaozhou Bay is a semi-enclosed coastal embayment located on the east coast of China and adjacent to the Yellow Sea (Figure 5). The area of Jiaozhou Bay is about 400 km^2 with an average water depth of 7 m. Jiaozhou Bay is surrounded by three cities: Qingdao, Jiaozhou and Jiaonan that have a combined population of over 7 million and is one of the most populated coastal regions in the east coast of China. Over the years, Jiaozhou Bay has been the most important resource for civil, industrial and commercial development for the seaport city of Qingdao. Jiaozhou Bay is also the home of Qingdao Port, located on the east side of the bay and adjacent to the city.

Qingdao Port is one of China's major trading ports, one of the top 7 x 10^8 ton ports in China, and one of the top 20 trading ports in the world. In the last ten years, the rapid urbanization along Jiaozhou Bay has affected the bay's ecosystem dramatically (Shen 2001; Wang et al. 2007). As a semi-enclosed system, the accumulation of pollutants discharged to the bay and deposition to the sediments has particular environmental concerns.

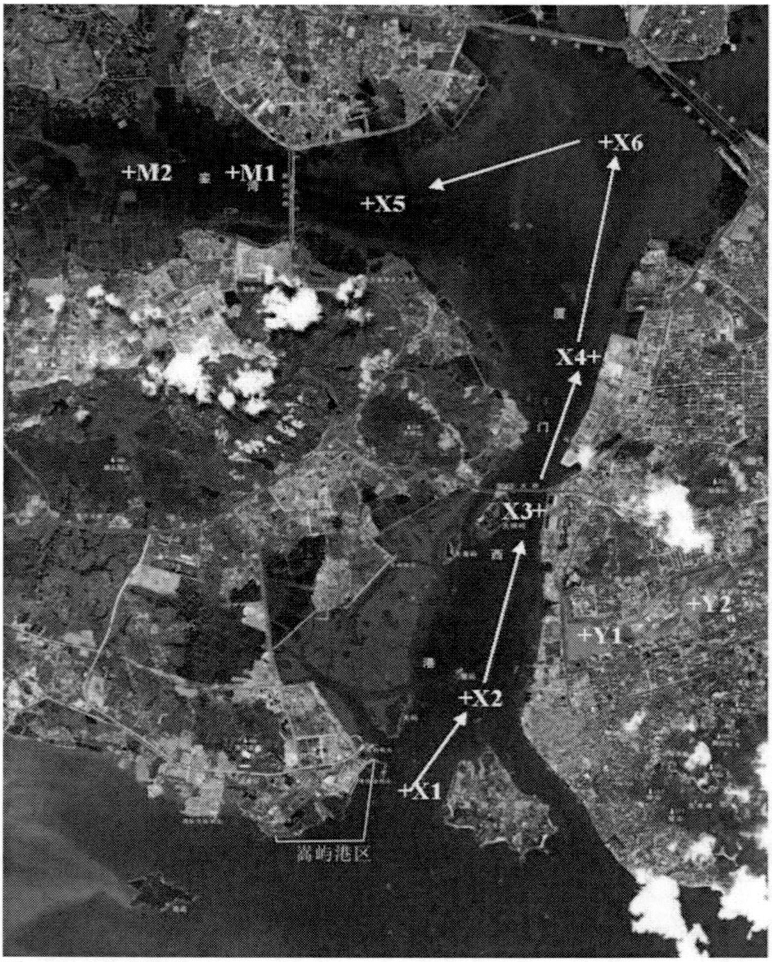

From Zhang e tal. 2007b.

Figure 4. Map showing the study area in the western Xiamen Bay, Xiamen, China.

Table 1. Heavy metal concentration in surface sediments (< 5cm) in selected urban estuaries and coastal areas in USA and China

Region		Ag (ppm)	Cd (ppm)	Cr (ppm)	Cu (ppm)	Hg (ppm)	Ni (ppm)	Pb (ppm)	Zn (ppm)	Reference
Hudson River Estuary, USA	mean±s.d.	2.23 ± 1.49	1.00 ± 0.56		77.9 ± 31.7			81.5 ± 33.5	182 ± 41	Feng et al. 1998a
	range	0.20 - 6.38	0.18 - 2.29		17.5 - 149			23.7 - 177	101 - 257	
New York Harbor, USA	mean±s.d.		4.6 ± 3.1		55 ± 30			92 ± 36	109 ± 62	Gibbs 1994
New York Harbor, USA	mean±s.d.		1.6 ± 0.5		135 ± 27			150 ± 31	265 ± 64	Chilrud 1996
Lower Passaic River, USA	mean±s.d.	6.37 ± 6.05	6.73 ± 4.11	198 ± 121		4.31 ± 2.33	61.5 ± 49.4	417 ± 338	681 ± 332	This Study
	range	0.62 - 42.3	0.32 - 29	8 - 860		0.1 - 12.4	7.1 - 369	4.4 - 2500	21 - 1900	
Yangtze River Intertidal Zone	mean±s.d.				56.6 ± 12.1			31.7 ± 6.1	233 ± 131	Feng et al. 2004
(near a major sewage outlet)	range				40.6 - 84.1			23.9 - 44.7	109 - 550	
Yangtze River Intertidal Zone	range		0.025 - 0.925	58.43 - 567	29.5 - 226			9.35 - 582	58.2 - 842	Xu et al. 1997
Xiamen Bay, China	mean±s.d.		0.33 ± 0.28	74.5 ± 25.5	44.0 ± 21.0		37.4 ± 12.1	54.0 ± 3.7	139 ± 39	Zhang et al. 2007b
	range		0.11 - 1.01	36.7 - 134	18.5 - 97.2		24.8 - 64.8	44.9 - 59.8	65 - 223	
Jiaozhou Bay, China	mean±s.d.		0.059 ± 0.027	22.1 ± 14.9	12.0 ± 7.5		13.2 ± 7.5	10.3 ± 4.1	31 ± 15	Wang et al. 2007
	range		0.030 - 0.110	5.0 - 51.0	4.2 - 28.0		2.7 - 23.0	5.1 - 18.0	12 - 58	
Continental Rocks		0.07	0.2	71	32		49	16	127	Martin and Whitfield, 1983

3. Result and Discussion

Table 1 summarizes metal contaminant concentrations in sediments in our selected areas in US and China. Significant spatial variations are seen in the study areas (Table 1). For comparison, natural levels of metal concentrations in rocks are also tabulated in Table 1. By comparison, it is seen that some metal concentrations in our study areas are above the natural levels, implying metal contamination. To properly evaluate anthropogenic heavy metal contamination in sediments, we can use enrichment factor (EF) of metals as an index. Mathematically, EF is a concentration ratio of observed metal to aluminum (or iron) in the sample of interest divided by the background metal/aluminum (or iron) concentration ratio and is expressed as (e.g., Covelli and Fontolan 1997; Feng et al. 1998a, 2004; Zhang and Liu 2002; Zhang et al. 2007b):

$$EF = \frac{\left(\frac{Me}{Al\,(\mathrm{Fe})}\right)_{Sample}}{\left(\frac{Me}{Al\,(\mathrm{Fe})}\right)_{Background}} \qquad (1)$$

where $(\mathrm{Me/Al(Fe)})_{\mathrm{Sample}}$ is the metal to Al (or Fe) ratio in the samples of interest; $(\mathrm{Me/Al(Fe)})_{\mathrm{Background}}$ is the natural background value of metal to Al (or Fe) ratio.

Advantage of using enrichment factor (EF) is that it can reflect the degree of metal contamination in the environment. This technique has been well applied in several studies to assess metal contamination in marine sediments (Windom et al. 1989, 1991; Zhang and Liu 2002; Feng et al. 1998a, 2004; Lu et al. 2005; Zhang et al. 2007b). The assessment criteria from Han et al. (2006) are listed in Table 2. In general, if an EF value is less than 2 (i.e., $EF < 2$), it suggests that the trace metals may be entirely from crustal materials or natural weathering processes, while a value of EF is greater than 2 (i.e., $EF>2$), it suggests that a significant portion of trace metal is delivered from non-crustal materials or non-natural weathering processes and a varying degrees of metal contamination exist depending on EF values (Sutherland 2000; Loska and Wiechuya 2003; Zhang and Liu 2002; Han et al. 2006).

3.1. Lower Hudson River - New York Harbor, USA

In the lower Hudson River estuary including New York Harbor, the concentrations of metal contaminants show a wide spatial variation. Feng et al. (1998a) reported that metal concentrations in the Hudson River estuary sediments range from 0.20 to 6.38 ppm for Ag, 0.18 to 2.29 ppm for Cd, 18 to 149 ppm for Cu, 24 to 177 ppm for Pb and Zn from 101 to 257 ppm for Zn. (Table 1). Gibbs (1994) found that metal concentrations in the New York Harbor sediments were 4.6$\pm$3.1 ppm for Cd, 55$\pm$30 ppm for Cu, 92$\pm$36 ppm for Pb and 109$\pm$62 ppm for Zn (Table 1). Chillrud (1996) also reported the concentrations of metal contaminants in the New York Harbor, which are 1.6$\pm$0.5 ppm for Cd, 135$\pm$37 ppm for Cu, 150$\pm$31 ppm for Pb and 265$\pm$64 ppm for Zn (Table 1).

Results from the previous studies indicate that mechanisms and processes governing the heavy metal contaminant input and distributions in the harbor could be explained by their chemically particle-reactive characteristics and estuarine hydrodynamics (Ullman and Wilson 1998; Feng et al. 2002). . When these metal contaminants were discharged into the Hudson River from the sources, they were scavenged onto the surface of suspended particles in water column, settled down to sediments along with the particles, and then resuspended and redistributed in the estuary (Hines et al. 1984; Windom et al. 1991; Feng et al. 1999a). Feng et al. (1999b) estimated that the residence times of ^{234}Th- and ^{7}Be-carrying particles in the Hudson estuary range from <1 to 10 d and the sediment could be transported up to 10-15 km up- and down-estuary during a tidal cycle.

Based on contamination categories by Han et al. (2006) (Table 2), metal enrichment factors in the Hudson River estuary sediment indicate that, the Hudson River estuary are generally contaminated by Ag, Cd, Cu, Pb and Zn and the contamination are more significant in some areas (Table 3).

Sediment contamination in the Hudson River estuary sediments has been studied by a number of workers since the 1970s (e.g., William et al. 1978; Olsen et al. 1978, 1981, 1984; Rohmann 1988; Bopp and Simpson 1989; Gibbs 1994; Hunt et al. 1994; Chillrud 1996; Hirschberg et al. 1996), in order to better understand the environmental quality of the Hudson River estuary, and the sources and factors controlling the sediments-associated contaminant distributions. Williams et al. (1978) found that Cu, Pb and Zn concentrations in the New York harbor sediments were high and anthropogenic metals discharged from dispersed contaminant sources were apparent to depths >60 cm in cores taken in this area. According to Olsen et al. (1984), deposition of suspended particles in the harbor removes virtually the entire riverborne load of particle-associated contaminants.

Along the Hudson River estuary, Gibbs (1994) found that the bottom sediment has a metal concentration maximum in the New York Harbor. Normalization of metal concentrations to Fe indicated a maximum level of pollution in the harbor (Gibbs 1994). Gibbs (1994) attributed the high metal concentrations in the harbor to local sources as well as sediment properties and hydrodynamics. Chillrud (1996) found that most metal contaminants such as Cd, Cu, Pb and Zn in New York Harbor sediments had declined by 50-90% from their maximum levels reached in the 1960s and 1970s, and the decreases seemed continuing during the early 1990s.

In an urban estuarine system, Ag can be referred to a good indicator of urban waster effluent input (Sañudo-Wilhelmy and Flegal 1992; Smith and Flegal 1993). In the Hudson River estuary, Feng et al. (1998a) found that anthropogenic Cd, Cu, Pb and Zn showed linear correlations with anthropogenic Ag, suggesting that Cd, Cu, Pb and Zn were introduced to the Hudson River estuary along with Ag from the urban sources and distributed and accumulated in the area. It is very likely that that a certain amount of heavy metal contaminants are transported and redistributed with sediments due to estuarine circulation (Geyer 1993; Feng et al. 1999a, 1999b). These studies show that the Hudson River estuary is a heavily contaminated estuary.

3.2. Lower Passaic River, New Jersey, USA

The Passaic River basin contains some of the most densely populated land and the most polluted section of the river in the nation (Bonnevie et al. 1992, 1993, 1994; Gillis, et al. 1993 a, 1993b, 1995; Huntley et al. 1993, 1995, 1997; Iannuzzi and Wenning 1995; Iannuzzi et al. 1995, 1997, 2002; Wenning et al. 1993a, 1993b, 1994; Feng et al. 2005; Onwueme and Feng 2006). In this study, we analyzed the existing large dataset from the previous federal and New Jersey state sampling programs between 1990 and 2000 (e.g., R-EMAP 1998) and found that the contaminant concentrations in the lower Passaic River sediments are 6.37±6.06 ppm (ranging from 0.62 to 42.3 ppm) for Ag, 6.73±4.11 ppm (ranging 0.32-29 ppm) for Cd, 198±121 ppm (ranging 8-860 ppm) for Cr, 4.31±2.33 ppm (ranging 0.1-12.4 ppm) for Hg, 61.5±49.4 ppm (ranging 7.1-369 ppm) for Ni, 417±338 ppm (ranging 4.4-2500 ppm) for Pb, and 681±332 ppm (ranging 21-1900 ppm) for Zn (Table 1). We also calculated metal enrichment factors in this study and found that metal enrichment factor for Ag is 2.9±2.2 (ranging 0.2 - 15.9); Cd, 10.6±6.9 (0.4 - 51.2); Cr, 6.5±4.2 (1.0 - 32.4); Hg, 10.6±6.1 (0.5 – 37); Ni, 2.1±1.7 (0.7 – 12); Pb, 20±24 (0.2 – 207); and Zn, 8.2±4.1 (1.0 – 28) (Table 3). Based on the contamination categories (Han et al. 2006; Table 2), the metal enrichment factor indicate that the lower Passaic River is moderately to significantly contaminated by these metals as a whole and the contamination in some areas are very high.

Like other urban watershed and river-estuary systems, discharges of a variety of contaminants through runoff, river tributaries and industrial wastewater outfalls are problematic in coastal urban areas (Brosnan et al. 1994; Sañudo-Wilhelmy and Flegal 1994; Feng et al. 1998a, 2004; Zhang et al. 2007b). The lower Passaic River system is not an exception. In 1970, the United States Environmental Protection Agency (USEPA) declared the Passaic River the "second most polluted river in America," behind only the Cuyahoga, which had caught fire in 1969 (Peet and Johnson 1996), and the most chemically polluted estuary in the nation (NOAA 1984). In an earlier study, Feng et al. (2005) found that the Harrison Reach sediment in the lower Passaic River has the highest As, Cr, Hg, Pb and Zn concentrations. As there were and are numerous of industrial manufactures situated on the both side of the lower Passaic River, they could have been potential sources of these contaminants to the Passaic River (Bonnevie et al. 1992, 1993, 1994; Crawford et al. 1995; Gills et al. 1993a, 1993b; Huntley et al. 1997).

This is especially true in the past when the discharge of various contaminants to the Passaic River system was uncontrolled and unlimited (Huntley et al. 1995; Iannuzzi and Wenning 1995; Iannuzzi et al. 1997; Meyserson et al. 1981).

Table 2. Contamination categories based on enrichment factor values (Han et al. 2006)

EF value	Category
EF<2	deficiency to minimal enrichment
EF=2–5	moderate enrichment
EF=5–20	significant enrichment
EF=20–40	very high enrichment
EF>40	extremely high enrichment

3.3. Yangtze River Estuary, China

In the Yangtze River intertidal zone, sediments near a point source of contaminants (e.g., a major sewage outlet) can preserve a record of contamination from the source and, therefore, metal concentration in the sediment could reflect an environmental impact due to urbanization. Feng et al. (2004) conducted a preliminary study in high, middle and low tidal flats of the Yangtze River estuary intertidal zone. Three short sediment cores (<20 cm) were collected in the high, middle and low tidal flats in the Yangtze River Estuary near the Southern (Nanqu) Sewage Outlet, one of the three largest sewage outlets in Shanghai, China. They found that metal concentrations in surface sediments (<5 cm) are 56.6±12.1 ppm (ranging 40.6 - 84.1 ppm) for Cu, 31.7±6.1 ppm (ranging 23.9 - 44.7 ppm) for Pb and 233±131 ppm (ranging 109 - 550 ppm) for Zn (Table 1). In an earlier study in the Yangtze River estuary intertidal zone, Xu et al. (1997) found that Cd concentration in the sediments ranges from 0.025 to 0.925 ppm; Cr, 58.4 - 567 ppm; Cu, 29.5 - 226 ppm; Pb, 9.35 – 582 ppm; and Zn 58.2 – 842 ppm. Feng et al. (2004) also calculated the metal enrichment factor based on their study, which are 3.2±0.6, 1.6±0.3 and 4.7±2.4 for Cu, Pb and Zn, respectively with a range of 2.5 – 4.6, 1.3 – 2.1 and 2.4 - 10.8, respectively (Table 3).

Although the sediment in Yangtze River intertidal zone is mainly from Yangtze River upper stream sediment discharge (4.7×10^{8} tons y^{-1}, Chen 1998), our early preliminary study (Feng et al. 2004) in a localized area indicate that the environmental impact due to urbanization and industrial development, as shown by this localized intertidal zone, is significant as reflected by their statistical analysis that time has a significant influence on metal concentrations whereas grain size has minor influence. The results from Feng et al. (2004) show that the intertidal zone near the sewage outlet is moderately contaminated by Cu and Zn, and, to less extent, contaminated by Pb based on Han et al. (2006) (Table 2). Feng et al. (2004) also indicate that the metal contamination is especially true in surface sediments due to environmental impact in recent years. The high EF values (EF>2, Table 3) support the conclusion that anthropogenic input of the metals is mainly due to urbanization and economic development in Shanghai metropolitan area happened in the past 10-20 years.

Table 3. Heavy metal enrichment factors (EF)in surface sediments (< 5cm) in selected urban estuaries and coastal areas in USA and China

Region		Ag	Cd	Cr	Cu	Hg	Ni	Pb	Zn	Reference
Hudson River Estuary, USA	mean±s.d	9.4 ± 5.4	8.8 ± 4.3		5.7 ± 2.0			5.8 ± 2.1	2.3 ± 0.3	Feng et al. 1998a
	range	0.8 - 23	2.0 - 18		1.5 - 10			1.9 - 12	1.5 - 3.1	
Lower Passaic River, USA	mean±s.d	2.9 ± 2.2	10.6 ± 6.9	6.5 ± 4.2		10.6 ± 6.1	2.1 ± 1.7	20 ± 24	8.2 ± 4.1	This Study
	range	0.2 - 16	0.4 - 51	1.0 - 32		0.5 - 37	0.7 - 12	0.2 - 207	1.0 - 28	
Yangtze River Intertidal Zone	mean±s.d				3.2 ± 0.6			1.6 ± 0.3	4.7 ± 2.4	Feng et al. 2004
	range				2.5 - 4.6			1.3 - 2.1	2.4 - 11	
Xiamen Bay, China	mean±s.d		1.5 ± 1.2	1.0 ± 0.5	1.3 ± 0.6		0.7 ± 0.3	3.1 ± 0.4	1.0 ± 0.2	Zhang et al. 2007b
	range		0.6 - 4.1	0.6 - 2.2	0.8 - 2.7		0.4 - 1.5	2.6 - 3.8	0.8 - 1.4	
Jiaozhou Bay, China	mean±s.d		2.7 ± 2.6	1.8 ± 0.4	2.4 ± 1.0		1.6 ± 0.4	4.9 ± 3.0	1.7 ± 0.9	Wang et al. 2007
	range		1.0 - 7.6	1.4 - 2.7	1.6 - 5.0		1.2 - 2.4	2.5 - 12	1.0 - 3.9	

3.4. Western Xiamen Bay, China

In western Xiamen Bay, field samplings for surface sediments (<5 cm) were conducted in December 2004 and July 2005, respectively, in order to examine if there is a significant temporal variation of metal concentrations and distribution. The sampling stations in western Xiamen Bay, Maluan Bay and Yuandang Lagoon are shown in Figure 4.

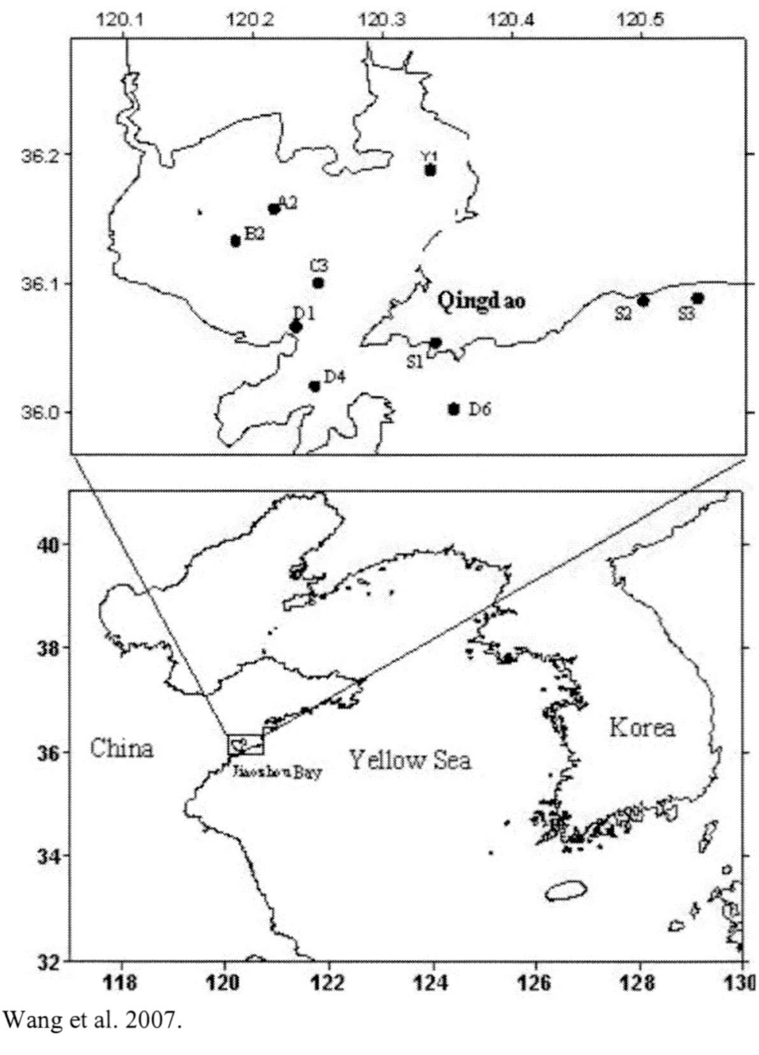

From Wang et al. 2007.

Figure 5. Map showing the study area in Jiaozhou Bay, China.

In a recent metal contamination study in western Xiamen Bay, China, Zhang et al. (2007b) found that metal concentrations in the surface sediments from the two samplings (December 2004 and July 2005, respectively) show a general consistency and the concentrations vary from 0.11-1.01 ppm for Cd, 37-134 ppm for Cr, 19-97 ppm for Cu, 25-65 ppm for Ni, 45-60 mg ppm for Pb and 65-223 ppm for Zn. They indicate that metal contaminant sources to the western Xiamen Bay and adjacent areas should include land-based point and non-point input, riverine/stream discharge and atmospheric fallout. Studies on atmospheric inputs to China sea have become a new focus of biogeochemical cycle in the past few years (e.g., Gao et al. 2002). However, this information of atmospheric fallout to the study area is missing based on our literature review. As the western Xiamen Bay area is within the coastal region, we expect the land-based influence should be more predominant.

The metal enrichment factors in Xiamen Bay sediment from Zhang et al. (2007) show that Cu enrichment factor, EF (Cu), ranges from 0.8 to 2.7, EF (Pb) from 2.6 to 3.8, and EF (Zn) from 0.8 to 1.4, EF (Cd) from 0.6 to 4.1, EF (Cr) from 0.6 to 2.2, and EF (Ni) from 0.4 to 1.5 (Table 3). Comparing with the metal contamination categories (Table 2), the metal enrichment factors (Table 3) indicate moderate Pb contamination in sediment appears everywhere in western Xiamen Bay, Ni and Zn contaminations are not significant at present, and moderate Cu, Cd, and Cr contaminations exist in some areas in western Xiamen Bay.

The results show that with the creation of Xiamen Economic Special Zone and establishment of industrial enterprises, the environmental impact has been seen as demonstrated by metal contamination in western Xiamen Bay and adjacent areas, which is becoming an environmental problem. Zhang et al. (2007b) indicated that although metal (Cu, Pb, Zn, Ni, Cd and Cr) concentrations in the surface sediments meet the Chinese National Standard of Marine Sediment Quality Criteria (CSTBS, 2002), metal enrichment factors (EF) and other index confirm the certain extent of metal contamination in the western Xiamen Bay and adjacent areas and that point source input of some specific metal contaminant(s) has caused significant sediment contamination in the local area.

3.5. Jiaozhou Bay, China

In our another study (Wang et al. 2007), sediments were collected from 10 sites in and outside Jiaozhou Bay (Figure 5) using an Ekman style stainless steel grab sampler (0.1 m^3). Wang et al. (2007) found that concentrations of toxic heavy metals ranged from 0.03 to 0.11 ppm for Cd, 5 to 51 ppm for Cr, 4.2 to 28 ppm for Cu, 2.7 to 23 ppm for Ni, 5.1 to 18 ppm for Pb, and 12 to 58 ppm for Zn. The distribution of heavy metal contaminants in the surface sediments showed significant spatial variations among the stations due to variability in metal source input.

They found strong correlations between metal contaminants (Cu, Cr, Ni, Pb and Zn) and Al in the sediments (r^2 = 0.892 - 0.973), indicating that these heavy metals were strongly associated with the clay mineral phases in the sediments since Al is one of the most abundant elements in clay minerals and the variation of Al concentration usually suggests differences in particle grain size and Al-Si clay mineral content.

Wang et al. (2007) also used enrichment factor (EF) as an index to examine the extent of metal contamination in the sediments of Jiaozhou Bay. Based on the contamination categories (Han et al. 2006; Table 2), the metal EF values indicate that the sediments in Jiaozhou Bay have been contaminated by Cd, Cr, Cu, Ni, Pb and Zn to various degrees that, with a wide variation, Cd, Cu and Pb show a moderate enrichment in the sediments on average while Cr, Ni and Zn show a less contamination (Table 3). Lead (Pb) contamination is relatively more significant.

Although the sediment quality guidelines (SQDs) (Long et al. 1995, 1998) are useful indicators for assessing sediment contamination, Wang et al. (2007) show that the level of contamination may not be properly evaluated by comparing metal concentrations only without considering sedimentary mineralogy and geochemistry. For example, enrichment factor of metals is a useful indicator for assessment of metal contaminations in sediments. Based on their results, Wang et al. (2007) indicate that sediment grain size and organic matter play important roles in controlling the distribution of the heavy metals in surface sediments of Jiaozhou Bay and the major inputs of metal contamination to Jiaozhou Bay are from land-based anthropogenic sources such as discharges of industrial wastewater and municipal sewage and run-off. The elevated heavy metal concentrations in Jiaozhou Bay sediments could have significant impact on the bay's ecosystem. With the rapid economic development and urbanization around Jiaozhou Bay, coastal management and pollution control should focus on these contaminant sources, and continuing monitoring studies of heavy metal contamination within the bay is needed.

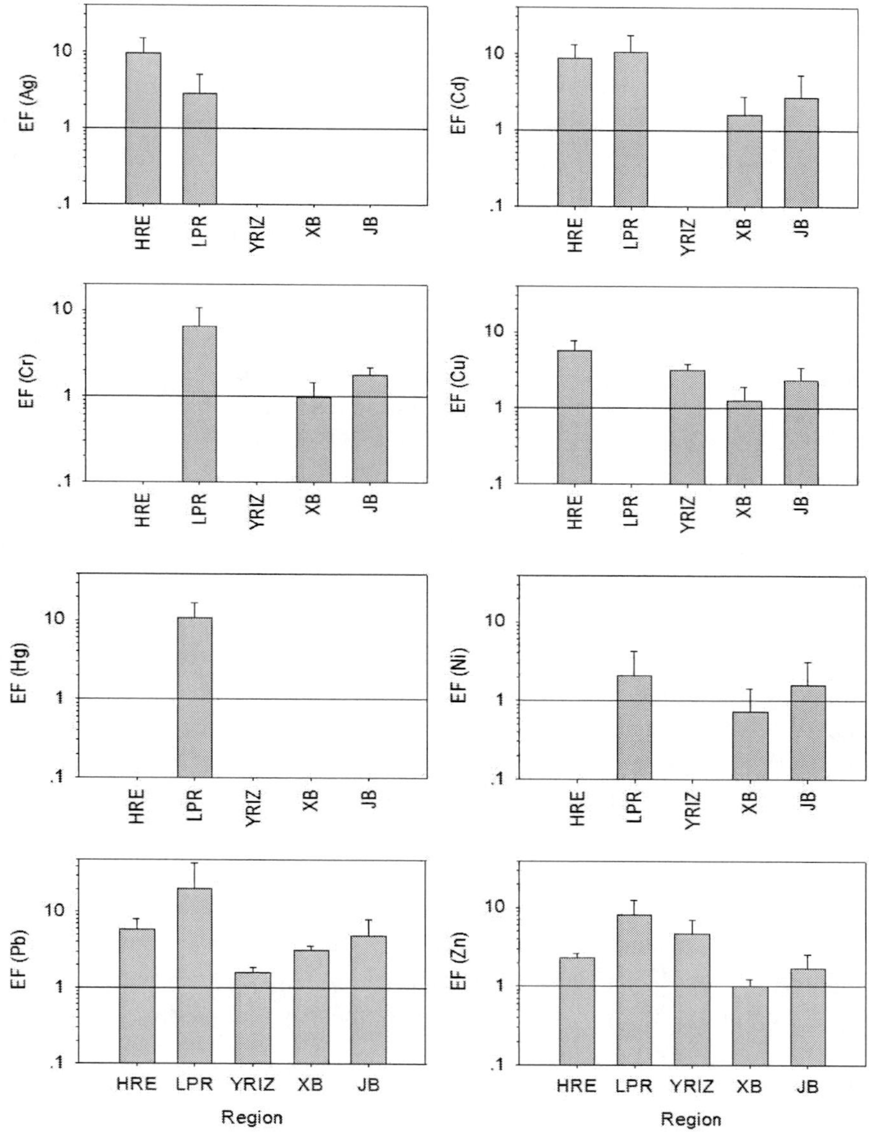

Figure 6. Comparison of metal enrichment factors (EF) in our selected areas. The predominant metal contaminants vary from one place to another due to the regional industrial and economic settings. However, varying extent of metal contamination are seen in these regions. The lower Hudson River estuary is designated as HRE, the lower Passaic River as LPR, Yangtze River intertidal zone as YRIZ, Xiamen Bay as XB, and Jiaozhou Bay as JB.

4. Summary

In studying metal pollution in sediments, the extent of metal contamination can be categorized using the metal enrichment factor as suggested by Han et al. (2006). In this study, we compared the metal enrichment factors in different coastal regions (Figure 6). As shown in Figure 6, although the predominant metal contaminants vary from one place to another depending on the local industrial and economic setting, there is no exception to find that varying extent of metal contamination does exist in these regions. With the rapid industrialization and economic development in coastal regions, if lack of proper environmental protection procedures and regulations, heavy metals could be continuously introduced to estuarine and coastal environment through rivers, runoff and land-based point sources where metals are produced as a result of metal refinishing byproducts. Therefore, heavy metal contamination will have been still an environmental problem facing our society. The sediment quality guidelines, criteria, residential and non-residential standards, etc. established by federal and state environmental protection agencies in various countries are critical and useful for protecting human environment and assessing sediment/soil environmental quality. However, a multiple approach should be applied to properly evaluate and reflect the environmental status of an area and ensure the effectiveness in implementation of these regulatory requirements. As to the significantly polluted areas, remediation and restoration actions are needed to recover the system.

For the future work, we suggest that an in-depth understanding of the source, transport and fate of these contaminants be necessary for effective environmental management and ecosystem restoration. With the rapid urbanization and economic development in coastal areas, continuing monitoring and ecological studies of heavy metal pollution are certainly still needed.

Acknowledgments

We would like to thank Libertad Urena and Jackie Kinney for their assistance in preparation of this manuscript.

REFERENCES

Abood, K. A. (1978) Hudson River hydrodynamic and water quality characteristics. Prospectives on the New York Bight Symposium. Williamsburg, VA, 5-10 Nov. 1978.

Bonnevie N. L., Wenning R.J. and Gunster, G.D. (1992) Lead contamination in surficial sediments from Newark Bay, New Jersey. *Environmental International*, 18, 497-508.

Bonnevie N.L., Wenning R.J. and Huntley S.L. (1993) Distribution of inorganic compounds in sediments from three waterways in Northern NJ. *Environmental Contamination and Toxicology*, 51, 672-680.

Bonnevie, N. L., Huntley, S. L., Found, B. W. and Wenning, R. J. (1994) Trace metal contamination in surficial sediments from Newark Bay, New Jersey. *Science of Total Environment*, 144, 1-16.

Bopp, R. F. and Simpson, H. J. (1989) Contamination of the Hudson River - The sediment record. In: Contaminated Marine Sediments Assessment and Remediation. pp. 401-416.

Bothner, M.H., Buchhltz ten Brink, M. and Manheim, F.T. (1998) Metal concentrations in surface sediments of Boston Harbor – Changes with time. *Marine Environmental Research* 45, 127-155.

Brosnan, T. M., Stubin, A. I., Sapienza, V. and Ren, Y. G. (1994) Recent changes in metals loadings to New York Harbor from New York City water pollution control plants. In: Hazardous and Industrial Wastes. C. P. Huang (eds.). pp. 657-666.

Chen, X.Q. 1998. Changjiang (Yangtze) River Delta, China. *Journal of Coastal Research*, 14, 838-858.

Chen, Z., Kostaschuk, R. and Yang, M. (2001) Heavy metals on tidal flats in the Yangtze Estuary, China. *Environmental Geology* 40, 742-749.

Chillrud, S. N. (1996) Transport and fate of particle associated contaminants in the Hudson River basin. Ph. D. dissertation, Columbia University, Palisades, New York. 277 p.

Cooper, J. C., Cantelmo, F. R. and Newton, C. E. (1988) Overview of the Hudson River Estuary. *American Fisheries Society Monograph*, 4, 11-24.

Covelli, S. and Fontolan, G. (1997) Application of normalization procedure in determining regional geochemical baseline. *Environmental Geochemistry* 30, 34-45.

Crawford, D. W., Bonnevie, N. L. and Wenning, R. J. (1995) Sources of pollution and sediment contamination in Newark Bay, New Jersey, *Ecotoxicology and Environmental Safety*, 30, 85-100.

CSBTS (China State Bureau of Quality and Technical Supervision) (2002) The People's Republic of China National Standards GB 18668-2002 - Marine Sediment Quality. 10 pp. (In Chinese).

Dai, W.M. and Gu, Y.Z. (1990) Accumulation of heavy metal elements in the sediments near Xiqu and Nanqu sewage outlets, Shanghai. Shanghai *Environmental Science*, 9, 38–40 (in Chinese).

Feng, H., Cochran, J. K., Lwiza, H., Brownawell , B. and Hirschberg, D. J. (1998a) Distribution of Heavy Metal and PCB contaminants in the sediments of an urban estuary: the Hudson River. *Marine Environmental Research*, 45, 69-88.

Feng, H., Cochran, J. K., Hirschberg, D. J. andWilson, R. E. (1998b) Small-scale spatial variations of natural radionuclide and trace metal distributions in sediments from the Hudson River Estuary. Estuaries, 21, 263-280.

Feng H., Cochran, J.K. and Hirschberg, D.J. (1999a) ^{234}Th and ^{7}Be as tracers for the sources of particles to the turbidity maximum zone of the Hudson River estuary. Estuarine, Coastal and Shelf Science, 49, 629-645.

Feng, H., Cochran, J. K., and Hirschberg, D. J. (1999b) ^{234}Th and ^{7}Be as tracers for the transport and dynamics of suspended particles in a partially mixed estuary. Geochimica et Cosmochimica Acta, 63, 2487-2505.

Feng, H., Cochran, J.K. and Hirschberg, D.J. (2002) Mechanisms and behavior of metal contaminants over the course of a tidal cycle in the turbidity maximum zone of the Hudson River estuary. Water Research, 36, 733 – 743.

Feng, H., Han X., Zhang, W. and Yu, L. (2004) A preliminary study of heavy metal contamination in Yangtze River intertidal zone due to urbanization. Marine Pollution Bulletin. 49, 910–915.

Feng, H., Onwueme, V., Jaslanek, W. J., Stern, E. A., and Jones, K.W. (2005) Application of geographical information system (GIS) to lower Passaic River sediment pollution study, New Jersey, USA. In: Urban Dimensions of Environmental Change —— Science, Exposure, Policies, And Technologies. (eds. H. Feng, L. Yu and W. Solecki). p p. 275-282. Science Press USA Inc., Monmouth Junction, New Jersey, USA.

Finley, B., Wenning, R. J., Ungs, M. J., Huntley, S.L. and Paustenbach, D. J. (1990) PCDDs and PCDFs in surficial sediments from the lower Passaic River and Newark Bay, p.409-414, In: Short Papers from the 10th International Meeting, Dioxin '90, Volume 1.

Forstner, U. and Wittman, G.T.W., (1983). Metal pollution in the aquatic environment, 2$^{nd.}$ Ed., Verlag Publishers. New York. 18-19.

Gao, H.-W., Zhang, Y.-J, and Zhang, K. (2002) Atmospheric inputs of pollutants to the sea and their effects on marine envioronment and ecosystem. Advance in earth Sciences, 17, 326-330. (In Chinese with English abstract.)

Geyer, R.W. (1993) The importance of suppression of turbulence by stratification on the estuarine turbidity maximum. Estuaries, 16, 113-125.

Gibbs, R. J. (1994) Metals in the sediments along the Hudson River Estuary. *Environmental International* 20, 507-516.

Gillis, C. A., Gunster, D. G., Bonnevie, N. L., Abel. T. B. andWenning, R. J. (1993a) Petroleum and hazardous chemical spills in Newark Bay, New Jersey, USA from 1982 to 1991. *Environmental Pollution*, 82, 245-253.

Gillis, C.A., Bonnevie, N.L. and Wenning, J. (1993b) Mercury contamination in the Newark Bay Estuary. *Ecotoxicology and Environmental Safety*, 25, 214-226.

Gillis, C. A., Bonnevie, N. L., Su, S. H., Ducey, G. J., Huntley, S. L. and Wenning, R. J. (1995) DDT, DDD, DDE contamination of sediment in Newark Bay Estuary, New Jersey. *Archives of Environmental Contamination and Toxicology*, 28, 85-92.

Goldberg, E.D., Griffin, J.J., Hodge, V., Koid, M.and Windom, H.L. (1979) Pollution history of the Savannah River estuary. *Environmental Science and Technology*, 13, 588-594.

Han, Y.M., Du, P.X., Cao, J.J. and Posmentier E.S. (2006) Multivariate analysis of heavy metal contamination in urban dusts of Xi'an, Central China. *Science of the Total Environment*, 355, 176– 186.

Hines, M. E., Lynons, W., Amstrong, P. B., Orem, W. H., Spencer, M. J., Gaudette, H. E. and Jones, G. E. (1984) Seasonal metal remobilization in the sediments of Great Bay, New Hampshire. *Marine Chemistry*, 15, 173-187.

Hirschberg D. J., Chin P., Feng H. and Cochran J. K. (1996) Dynamics of sediment and contaminant transport in the Hudson River Estuary: evidence from sediment distributions of naturally occurring radionuclides. Estuaries, 19, 931-949.

Hunt, C. D., West, D. A.and Lewis, D. A. (1994) Trace metals concentrations in New York/New Jersey Harbor. In: Hazardous and Industrial Wastes. (ed. C. P. Huang). pp. 691-698.

Huntley S.L., Bonnevie N.L., Wenning R.J. and Bedbury H. (1993) Distribution of polycyclic aromatic hydrocarbons (PAHs) in three northern New Jersey waterways. *Bulletin of Environmental Contamination and Toxicology*, 51, 865-872.

Huntley S.L., Bonnevie N.L. and Wenning R.J. (1995) Distribution of polycyclic aromatic hydrocarbons and petroleum hydrocarbon contamination in sediment from Newark Bay estuary, New Jersey, *Archives of Environmental Contamination and Toxicology*, 28, 93-107.

Huntley, S. L., Iannuzzi, T. J., Avantaggio, J. D., Carlson -Lynch, H., Schmidt, C. W., and Finley, B. L. (1997) Combined Sewer Overflows as Sources of Sediment Contamination in the Lower Passaic River, New Jersey, II, Polychlorinated Dibenze-p-dioxins, Polychlorinated Dibenzofurans, and Polychlorinated Biphenyls, *Chemosphere* 34:233-250.

Iannuzzi, T. J. and Wenning, R. J. (1995) Distribution and Possible Sources of Total Mercury in Sediments from the Newark Bay Estuary, New Jersey, *Bulletin of Environmental Contamination and Toxicology*, 55:901-908.

Iannuzzi T. J., Huntley S. L., Finley B. L. and Wenning R. J. (1995) Distribution of possible sources of polychlorinated biphenyls in dated sediments from Newark Bay estuary, New Jersey. *Archives of Environmental Contamination and Toxicology*, 28: 108-117.

Iannuzzi, T. J., Huntley S.L., Schmidt, C. W., Finley, B. L., McNutt, R. P. and Burton, S. J. (1997) Combined Sewer Overflows (CSOs) as Source of Sediment Contamination in the Lower Passaic River, New Jersey. 1. Priority Pollutant and Inorganic Chemicals, *Chemosphere* 34:213-231.

Iannuzzi, T. J., Ludwig, D. F., Kinnell, J. C., Wallin, J. M., Desvousges, W. H. and Dunford, R. W. (2002) A Common Tragedy: History of an Urban River, Amherst Scientific Publishers.

Klinkhammer, G. P. and Bender, M. L. (1981) Trace Metal Distributions in the Hudson River Estuary. Estuarine, *Coastal and Shelf Science*, 12, 629-643.

Langston, W.J. (1990) Toxic effects of metals and the incidence of metals pollution in marine ecosystems. In: Furness, R.W., Rainbow, P.S. (Eds.), *Heavy Metals in the Marine Environment*. CRC Press, Boca Raton.

Long, E.R., MacDonald, D.D., Smith, S.L. and Calder, F.D. (1995) Incidence of adverse biological effects within ranges of chemical concentrations in marine and estuarine sediments. *Environmental Management* 19, 81-97.

Long, E.R., Field, L.J. and MacDonald, D.D. (1998) Predicting toxicity in marine sediments with numerical sediment quality guidelines. *Environmental Toxicology and Chemistry* 17, 714-727.

Loska, K. and Wiechuya, D.(2003) Application of principle component analysis for the estimation of source of heavy metal contamination in surface sediments from the Rybnik Reservoir. *Chemosphere*, 51, 723– 33.

Lu, X.Q., Werner, I. and Young, T.M. (2005) Geochemistry and bio-availability of metals in sediments from northern San Francisco Bay. *Environmental International* 31, 593-602.

Martin, J.-M. and Whitfield, M. (1983) The significance of the river input of chemical elements to the ocean. In: Trace Metals in Sea Water, C. S. Wong, E. Boyle, K. W. Bruland, J. D. Burton and E. D. Goldberg (Eds), Plenum Press, New York, pp.265-296.

Meyerson, A. L., Luther, G. W. III, Krajewski, J. and Hires, R. I. (1981) Heavy Metal Distribution in Newark Bay Sediments, *Marine Pollution Bulletin* 12(7): 244-250.

NJDEP [New Jersey Department of Environmental Protection]. (1987) Passaic River Water Quality Management Study, NJDEP, Division of Water Resources.

NOAA [National Oceanic and Atmospheric Administration]. (1984) A Geochemical Assessment of Sedimentation and Contaminant Distribution in the Hudson-Raritan Estuary, Rockville, MD: NOAA, National Ocean Service.

Olsen, C. R., Simpson, H. J., Bopp, R. F., Williams, S. C., Peng, T. H., and Deck, B. L. (1978) A geochemical analysis of the sediments and sedimentation in the Hudson Estuary. *J. Sediment. Petrol.*, 48, 401-418.

Olsen, C.R., Simpson, H.J. and Trier, R.M. (1981) Plutonium, radiocesium and radiocobalt in sediments of the Hudson river estuary. *Earth and Planetary science Letters*, 55, 377-392.

Olsen, C.R., Larsen, I.L., Brewster, R.H., Cutshall, N.H., Bopp, R.F. and Simpson, H.J. (1984) A geochemical assessment of sediment of sedimentation and contaminant distributions in the Hudson-Raritan Estuary. NOAA Technical Report NOS OMS 2. 101 pp.

Onwueme V. and Feng H. (2006) Risk characterization of contaminants in Passaic River sediments, New Jersey. *Middle States Geographer*, 39, 13-25.

Peet J., and Johnson T. (1996) The Forgotten River, Newark Star-Ledger, Three-Part Series. Property Owners. Chicago, IL, University of Chicago Press Publishing.

R-EMAP (Regional Environmental Monitoring and Assessment Program). (1998) Final Report: Sediment Quality of the NY/NJ Harbor System.

Rohmann, S. O. (1988) Hudson River pollution: comparing point and nonpoint source emissions of hazardous chemicals. In: Symposium on Coastal Water Resources. American Water Resources Association. W. L. Lyke and T. J. Hoban (eds.). pp. 383-391.

Santos, I.R., Silva-Filho, E. V., Schaefer, C. E., Albuquerque-Filho, M. R. and Campos, L. S. (2005) Heavy metal contamination in coastal sediments and soil near the Brazilian Antarctic Station, King George Island. Marine Pollution Bulletin 50, 185-194.

Sañudo -Wilhelmy, S. and Flegal, A. R. (1992) Anthropogenic silver in the southern California Bight: A new tracer of sewage in coastal waters. *Environmental Science and Technology*, 26, 2147-2151.

Sañudo-Wilhelmy S.A. and Flegal A.R. (1994) The fate of metals in the Southern California Bight. *Environmental Science and Technology*, 26, 2147-2151.

Shen, Z.-L. (2001) Historical changes in nutrient structure and its influence on phytoplankton composition in Jiaozhou Bay. Estuarine, *Coastal and Shelf Science* 52, 211-224.

Smith, G. J. and Flegal, A. R. (1993) Silver in San Francisco Bay estuarine waters. Estuaries, 16, 547-558.

Sutherland, R. A. (2000) Bed sediment-associated trace metals in an urban stream, Oahu, Hawaii. Environmental Geology, 39, 611– 27.

Tierra Solutions, Inc. (2003) Executive Summary: Passaic River Study Area Preliminary Findings.

Trefry, J. H. and Shokes, R. F. (1981) History of heavy metal inputs to Mississippi Delta sediments. *Marine Environmental Pollution*, 2, 193-208.

Ullman, D. S. and Wilson, R. E. (1998) Model parameter estimation from data assimilation modeling: Temporal and spatial variability of the bottom drag coefficient. *Journal of Geophysical Research*, 103, 5531-5549.

Underwood, E.J. (1971) Trace Elements in Human and Animal Nutrition (3rd Editieon). Academic Press, Inc., New York.

Valette-Silver, N. J. (1993) The use of sediment cores to reconstruct historical trends in contamination of estuarine and coastal sediments. Estuaries, 16, 577-588.

Walker, W. J., McNutt, R. P. and Maslanka, C. K. (1999) The Potential Contribution of Urban Runoff to Surface Sediments if the Passaic River: Source and Chemical Characteristics, *Chemosphere* 38:363-377.

Wang X-C., Feng H. and Ma, H-Q. (2007) An Assessment of Metal Contamination in Surface Sediments of Jiaozhou Bay, Qingdao, China. CLEAN - Soil, Air,Water. 35(1), 62-70.

Wenning J., Harris M., Paustenbach D.J. and Bedbury H. (1993a) Principal component analysis of potential sources of polychlorinated Dibenzo-p-Dioxin and Dibenzofuran residues in surficial sediments from Newark

Bay, New Jersey, *Archives of Environmental Contamination and Toxicology*, 24: 271-289.

Wenning J., Harris M., Finley B. Paustenbach D.J. and Bedbury H. (1993b) Application of pattern recognition technique to evaluate polychlorinated Dibenzo-p-Dioxin and Dibenzofuran distributions in surficial sediments from the Lower Passaic River and Newark Bay. *Ecotoxicology and Environmental Safety*, 25, 103-125.

Wenning R.J., Bonnievie N.L. and Huntley S.L. (1994) Accumulation of metals, polychlorinated biphenyls, and plocyclic hydrocarbons in sediment from lower Passaic River, New Jersey. Archives of Environmental Contamination and Toxicology, 27:64-81.

Williams, S. C., Simpson, H. J., Olsen, C. R. and Bopp, R. F. (1978) Sources of heavy metals in sediments of the Hudson River Estuary. *Marine Chemistry*, 6, 195-213.

Windom, H. L., Schropp, S. J., Calder, F. D., Ryan, J. D., Smith, R. G. Jr., Burney, L.C., Lewis, F. G. and Rawlinson, C. H. (1989) Natural trace metal concentrations in estuarine and coastal marine sediments of the southeastern United States. *Environmental Sciences and Technology*, 23, 314-320.

Windom, H., Byrd, J., Smith, R. Jr., Hungspreugs, M., Dharmvanij, S., Thumtrakul, W. and Yeats, P. (1991) Trace metal-nutrient relationships in estuaries. *Marine Chemistry*, 32, 177-194.

Wolfskill L. A. and McNutt R.P. (2000) An Environmental Study of the Passaic River and its Estuary, Steton Hall Law Review, 29:37-46.

Xu, S. Y., Tao, J., Chen, Z. L., Chen, Z. Y. and Lü, Q. R. (1997) Dynamic accumulation of heavy metals in tidal flat sediments of Shanghai. Oceanologia et Limnologia Sinica, 28, 509-515 (In Chinese with English abstract).

Zhang J. (1999) Heavy metal compositions of suspended sediments in the Changjiang (Yangtze River) estuary: Significance of riverine transport to the ocean. Continental Shelf Research 19, 1521-1543.

Zhang, J. and Liu, C. L. (2002) Riverine composition and estuarine geochemistry of particulate metals in China - Weathering features, anthropogenic impact and chemical fluxes. Estuarine, Coastal and Shelf Science, 54, 1051-1070.

Zhang L., Ye X., Feng H., Jing Y., Ouyang T., Yu X., Liang R. and Chen W. (2007) Heavy Metal Contamination in Western Xiamen Bay Sediments and Its Vicinity, China. Marine Pollution Bulletin, 54, 974–982.

Zhang, W., Yu, L., Hutchinson, S. M., Xu, S., Chen, Z. and Gao, X. (2001) China's Yangtze Estuary: I. Geomorphic influence on heavy metal accumulation in intertidal sediments. *Geomorphology*, 41, 195-205.

Zhang W., Yu L., Lu M., Hutchinson S. M. and Feng H. (2007) Magnetic approach to normalizing heavy metal concentrations for particle size effects in intertidal sediments in the Yangtze Estuary, China. *Environmental Pollution*, 147, 238-244.

INDEX

A

B

C

D

E

F

G

J

K

L

M

Q

R

S

T

U

V

W

X

Z